UAVs and Urban Spatial Analysis

Tony H. Grubesic • Jake R. Nelson

UAVs and Urban Spatial Analysis

An Introduction

Springer

Tony H. Grubesic
Geoinformatics and Policy Analytics
Lab (GPAL)
School of Information
University of Texas at Austin
Austin, TX, USA

Jake R. Nelson
Geoinformatics and Policy Analytics
Lab (GPAL)
School of Information
University of Texas at Austin
Austin, TX, USA

ISBN 978-3-030-35864-8 ISBN 978-3-030-35865-5 (eBook)
https://doi.org/10.1007/978-3-030-35865-5

This Springer imprint is published by the registered company Springer Nature Switzerland AG
The registered company address is: Gewerbestrasse 11, 6330 Cham, Switzerland

Preface

People make many assumptions about why authors write books. They assume the goal is to become rich and famous, touring the world on a lecture circuit, making appearances on late-night talk shows, and signing copies of books to adoring fans. Others assume writing a book equates to a vanity project, where the authors believe they have an important story to tell about their vast life experiences and/or intellectual insights. Fortunately for us, very little of that applies here. First, we will be lucky to sell six copies of this book. Second, we do think small unmanned aerial systems (sUAS) are important and that the future of urban spatial analysis will be strongly coupled with drones. Beyond that, our motivation for pursuing this book is a simple one. Drones are really cool. We like to fly them, we like the data they produce, we like making maps, and we like spatial analysis. If you like these things too, then this might be the book you are looking for.

The challenge in writing this book is that sUAS technology is moving so quickly that it is almost impossible to keep up. We have no doubt that by the time this book is published, sUAS will be more advanced, faster, better, stronger, etc. As authors, we find this somewhat frustrating, but it is the nature of the beast. Considering the speed at which the sUAS industry is moving, we want to encourage readers to dig deeper. Use this book as a foundation for learning more about sUAS technology, regulation, policy, and how all of it relates to (and can be used for) analyzing urban areas.

This book should appeal to a wide range of scholars, including geographers, planners, sociologists, criminologists, environmental scientists, and any other discipline that may, at least from time to time, require sUAS as part of their research portfolio. We also believe that this book would be ideal for instructors tasked with developing and teaching an introductory course on sUAS for urban analysis.

Our work is ongoing. We hope that readers enjoy this book as much as we liked writing it. Let's get flying!

Austin, TX, USA Tony H. Grubesic
Austin, TX, USA Jake R. Nelson
September 2019

Acknowledgements

We would like to express our thanks to a number of people. First, Dani Wallace (Arizona State University) and Alyssa Chamberlain (Arizona State University) humored our morning flight schedules and lots of trips to bug-infested locations to help us collect these data. We would also like to thank our awesome crew of graduate and undergraduate students at ASU, especially Jason Walker and Maribeth Hoyle, for hanging out in the heat, walking the neighborhood(s), and helping us coordinate our survey efforts. A special shout-out to Brian the dog, the guy who nicknamed our eBee "Hellfire," and the Unauthorized Bash Brothers Experience. Lastly, a quick hello to the dude in Norton Park who asked us to violate the federal statute 18 USC Section 1702.

Contents

Acronyms

AED	Automated external defibrillator
AGL	Above ground level
AMA	Academy of Model Aeronautics
ATC	Air traffic control
AUVSI	Association for Unmanned Vehicle Systems International
BVLOS	Beyond Visual Line of Sight
CCTV	Closed-circuit television system
CIR	Color infrared
COA	Certificate of authorization
DAA	Detect and avoid
DEM	Digital elevation model
DSM	Digital surface model
DTM	Digital terrain model
FAA	Federal Aviation Administration
FLIR	Forward-looking infrared
FoV	Field of view
GCP	Ground control point
GIS	Geographic Information System
GLONASS	Global Navigation Satellite System
GPS	Global Positioning System
GPT	General purpose technologies
GPU	Graphics processing unit
GSD	Ground sampling distance
GSV	Google Street View
IFR	Instrument flight rules
ISR	Intelligence Surveillance Reconnaissance
LiDAR	Light Detection and Ranging
LISA	Local Indicators of Spatial Association
MAUP	Modifiable areal unit problem
MEF	Maximum elevation figure
MSL	Mean sea level
NAD	North American Datum

NIR	Near infrared
NOTAM	Notice to Airmen
PPK	Post-processing kinematics
RGB	Red, Green, Blue color model
RINEX	Receiver Independent Exchange Format
RPIC	Remote Pilot in Command
RPV	Remotely piloted vehicle
RTK	Real-time kinematics
SfM	Structure from motion
SLAM	Simultaneous localization and mapping
sUAS	Small Unmanned Aerial System
sUAV	Small Unmanned Aerial Vehicle
UAV	Unmanned Aerial Vehicle
UTM	Universal Transverse Mercator
VFR	Visual flight rules
VLOS	Visible line of sight
VTOL	Vertical takeoff and landing
WH	Watt hours
ZCTA	ZIP Code Tabulation Areas

Chapter 1
Introduction to Small Unmanned Aerial Systems (sUAS) and Urban Spatial Analysis

Abstract There is a great deal of excitement regarding the use of small unmanned aerial systems for mapping and spatial analysis. The enthusiasm for this emerging technology is well deserved. No other data collection platform offers a similar combination of cost-effectiveness, quality, frequency, flexibility, extensibility, portability, safety, and reliability. The purpose of this chapter is to highlight both the promises and pitfalls of sUAS for urban spatial analysis, helping readers make a more informed decision on whether to include sUAS in their technology portfolio. We conclude this chapter by providing an organizational overview of the book, more clearly delineating the material to be covered in subsequent chapters.

Keywords General purpose technology · Low cost · High-quality imagery · Flexibility · Extensibility · Portability · Safety · Reliability · Temporal frequency · LiDAR · Mapping

1.1 Introduction

Drones have captured the public's imagination in many ways. Not only are they used as a plot device in many films (see *The Bourne Legacy* or *Syriana*), but they are also revolutionizing the way that commercials, television shows, and films are made. In addition to providing a bird's eye view of the action, Zacharek (2018) notes the significant cost advantages associated with their use in cinematography. Recent estimates suggest that the use of a manned helicopter can cost between \$20,000 and \$40,000 for a 10-h movie shoot, while the cost for a drone, operation crew, and insurance for the same 10-h shoot would be between \$4500 and \$13,000 per day (Zacharek 2018).

This is a substantial savings and because the quality of drone-generated footage is so good, the use of drones is drastically changing the workflow in many media productions. This is true for other industries as well, from agriculture (Anderson 2019; Tripicchio et al. 2015) to telecommunications (Giannini et al. 2016; Sydell 2017). Drones are seemingly everywhere, permeating every industry. It is also important to acknowledge that although the media favors the use of the term "drone,"

T. H. Grubesic, J. R. Nelson, *UAVs and Urban Spatial Analysis*,
https://doi.org/10.1007/978-3-030-35865-5_1

the aviation industry rarely uses it. There are several reasons for this. Consider, for example, the dictionary definition of a drone (MW 2019), "an unmanned aircraft or ship guided by remote control or onboard computers." While this is certainly accurate, it drastically underrepresents the complex *system of systems* that form the foundation of unmanned aerial systems (UAS). UAS consist of the flight platform (e.g., fixed wing, multi-rotor), onboard computers, receivers, transmitters, transceivers, multiplexers, controllers, proprioceptive sensors (e.g., accelerometer, vibration, global positioning system), exteroceptive sensors (e.g., range, contact, environmental temperature, humidity), electro-optical sensors (e.g., visual, ultraviolet, infrared), communication equipment (e.g., speaker and microphone), payload distribution systems (e.g., fertilizer sprayers), and many other types of interconnected hardware and software. The size of UAS can also vary substantially, but for the purposes of this book, we are generally focusing on small UAS (sUAS), which weigh under 55 lbs (~25 kg).

1.2 UAS as a General Purpose Technology

Given the ubiquitous nature of UAS and indicators suggesting the drone industry may be valued at $100 billion by 2020 (Goldman Sachs 2018), it is worth considering where UAS fit within the historical spectrum of disruptive technologies. For example, consider general purpose technologies (GPTs), which are defined by four key factors (Lipsey et al. 2005, 98): (1) it is a single, generic technology, recognizable as such over its whole lifetime, (2) initially has much scope for improvement and eventually comes to be widely used in the economy, (3) has many uses, and (4) generates many spillover effects. To date, there are only 24 recognized GPTs, all of which can generally be described as a *process*, *product*, or *organization*. For example, iron is a GPT that is classified as a product. Other GPTs include writing (process), the steam engine (product), factory systems (process), electricity (product), mass production (organization), the Internet (product), nanotechnology (product), artificial intelligence (process), and the airplane (product), among others.

Where the latter is concerned, it is tempting to directly connect airplanes to UAS. They both fly, so UAS must certainly be a GPT too, right? Not exactly. Consider the airplane in coordination with the Lipsey et al. (2005) framework. First, the airplane is a single generic technology, easily recognizable over its lifespan. Second, although it had much scope for improvement early on (WIPO 2015), it is now used widely across the global economy. Third, it has many uses, from transporting people and cargo, to its use in warfare. Finally, it generates many spillover effects (Kotha 2010; Niosi and Zhegu 2010), including the creation of completely new industries dedicated to supporting the continued development of aerospace technologies. Clearly, the airplane is a GPT.

While unmanned aerial systems share many characteristics with airplanes, this does not necessarily qualify them as GPTs. If anything, skeptics would argue that UAS are spillovers from the development of airplanes. This may be true, but to be fair, now consider UAS within the Lipsey et al. (2005) framework. First, UAS are a single generic technology, easily recognizable, even if their platforms adhere to several different forms (e.g., fixed wing, rotor, multi-rotor). Second, there is no doubt that UAS have much room for improvement—from endurance and reliability to speed and stability. This work is ongoing (Petritoli et al. 2018; Schenkelberg 2016). Third, as will be detailed throughout this book, UAS have many uses, from precision agriculture and mapping to cargo delivery and environmental sensing. Finally, are UAS associated with spillover effects? Although the UAS economy continues to develop, recent estimates suggest that by 2026 commercial drones will have an annual impact of \$31–\$46 billion on the US economy (Levick 2018). More importantly, although UAS technology is disruptive, in that it will displace less-efficient products and services, the UAS industry will also be constructive because it can create thousands of new jobs and contribute to multiple spillovers, rooted in both knowledge and technology (Levick 2018).

1.3 sUAS for Mapping and Urban Spatial Analysis

If there is one domain where sUAS technology is already disrupting the status quo, it is mapping and spatial analysis. As detailed by Robinson et al. (2005), a map is a "carefully designed instrument for recording, calculating, displaying, analyzing, and understanding the interrelation of things." More importantly, regardless of their intended function, all maps have two things in common: (1) maps represent locations and (2) maps represent attributes. Locations typically correspond to positions on the earth's surface (latitude and longitude) and are represented with x, y coordinates. Often, the coordinate pair is supplemented with a third value, z, which typically represents elevation. But, z could also be used to represent any attribute, from annual precipitation to the number of crimes that occurred at a given location. As detailed by Bishop and Grubesic (2016), these basic elements of geographic information can be used to derive a wide range of relational information from similarly referenced attributes, facilitating spatial analysis. For example, one could tabulate the differences in crime between two or more locations, or the relationship between crime and neighborhood poverty levels, education, or the geographic proximity of pawn shops to crime hot spots.

The question at hand, then, is why have sUAS been so disruptive to the mapping and spatial analysis industry over the past few years? There are eight key facets worth noting: (1) *low cost*, (2) *high quality*, (3) *frequency*, (4) *flexibility*, (5) *extensibility*, (6) *portability*, (7) *safety*, and (8) *reliability*.

First, as mentioned previously, entry-level drones, capable of basic mapping, are *low cost*. A recent industry survey by DroneDeploy suggests that 84% of drone mapping is conducted by sUAS that cost \$1500 or less (French 2017). This includes the wide range of DJI drones, such as the Phantom 4 and Phantom 4 Pro (v.2.0), with prices around \$1500. In short, there are few economic barriers for entry.

A second way sUAS are disrupting mapping and spatial analysis is the enviable combination of low entry costs and *high-quality* imagery, especially when compared to the alternatives. For example, as detailed by Manfreda et al. (2018), obtaining a new, high-resolution (50 cm/pixel) natural color image from a satellite, such as GeoEye-1, can cost up to $3000, per image. Conversely, one could opt for the use of a DJI Phantom Pro (v.2.0) ($1500), which offers a 20-megapixel camera capable of achieving 8 cm/pixel when flown at 300 feet, above ground level (AGL). At half the price and 4× the quality of the GeoEye-1 image, the benefits are obvious.

A third way that sUAS are disrupting mapping and spatial analysis is the temporal *frequency* in which data can be collected. As detailed by Grubesic et al. (2018), sUAS offers superior temporal frequency when compared to satellite and airborne alternatives. For example, DigitalGlobe notes that the WorldView 3 satellite, the most advanced panchromatic satellite sensor in the world, can visit any place on earth within 24 h. However, this is only partially true. Yes, at latitudes above 40°, it is possible to capture every patch of ground, daily. But, as the WorldView satellite moves closer to the equator at 0°, the overall landmass covered by the satellite drastically increases, and full, daily collections are not possible. While comparing the spatial coverage between an sUAS to the WorldView satellite is somewhat ridiculous, the WorldView will win every time. However, the important takeaway here is the notion of temporal coverage. An sUAS can be flown every day, even multiple times per day, over the same patch of land. This level of repeatability is limited only by the cost of fieldwork, capital investment costs for the sUAS, data storage, and associated computing/processing charges (Manfreda et al. 2018). In short, this level of temporal coverage is unmatched by any other mapping technology.

Small unmanned aerial systems are also massively *flexible*, in the sense that there are many different application areas in which they can be used. As noted earlier, this includes not only agriculture (Rokhmana 2015) and telecommunication (Orfanus and de Freitas 2014) but also urban planning (Grubesic et al., 2018), livestock management (Webb et al. 2017), search and rescue (Silvagni et al. 2017), law enforcement (Murphy and Cycon 1999), infrastructure inspection (Ellenberg et al. 2014), environmental monitoring (Lu et al. 2015), cargo transport (Maza et al. 2009), and 3D mapping (Nex and Remondino 2014), among others. There are very few technology platforms that can match this level of flexible application.

Most sUAS are also *extensible*. Although there are a handful of recreational, off-the-shelf sUAS that come equipped with standard payload/camera package that cannot be swapped out, commercial sUAS have a variety of payloads available. For example, consider the WingtraOne (URL: https://wingtra.com/drone/). Two different camera packages are available, including a Sony RXRII or a Sony QX1, as well as the MicaSense RedEdge-M (multispectral) and the FLIR Duo Pro thermal sensor. Similarly, the eBee X (https://tinyurl.com/yd349169) can fly with the S.O.D.A. 3D (which captures two oblique and one nadir photo each time), the Aeria X, Duet T, and the Parrot Sequoia+ (multispectral). In short, depending on user needs, commercial sUAS are highly extensible and can accommodate the appropriate payloads for a variety of applications.

Perhaps the most important (but often overlooked) characteristic of sUAS is their *portability*. There are many locations that are simply not accessible to vehicles but require fieldwork, mapping, and analysis. For example, in Arizona, there are hundreds of ecologically significant locations in the Sonoran Desert that require a hike (versus a drive) for access. This precludes the ability to use a truck-mounted LiDAR system or a larger UAS for analysis. Once again, consider the Ebee Classic, which weighs 1.52 lbs (0.69 kg) with the included camera and battery. There are very few scenarios where hiking into a field site with the eBee would be too difficult. Yes, one also needs to haul-in the telemetry unit, a computer and a few other items (e.g., additional batteries and a field repair kit), but this is easily done with a backpack. Because of this high-level of portability, there are very few locations where mapping and spatial analysis with sUAS are *not* possible.

Over the past several years, the *safety* technology embedded within sUAS has improved dramatically. On the hardware side, the use of GPS and GLONASS satellite navigation systems allow the sUAS to know where it is flying at all times. But more importantly, most entry-level commercial sUAS now include obstacle detection and collision sensors to avoid situations that might damage the drone, property, or people. Although obstacle avoidance sensors can take many forms (e.g., stereovision, ultrasonic, LiDAR, infrared), most sUAS use some combination of these detection technologies for obstacle avoidance. For example, the DJI Mavic 2 Pro includes forward, backward, downward, upward, and lateral vision sensors for omnidirectional obstacle detection. In this instance, the upward and downward sensors are infrared, while forward and the side system consists of vision sensors. Further, the built-in Mavic "terrain-follow" function uses height information gathered by the onboard ultrasonic system and uses its downward-facing cameras to keep the sUAS at a prescribed height above the ground. The DJI Mavic is not alone, Kespry, Yuneec, Wingtra, Sensefly, and many other manufacturers include similar systems.

On the software side, there have been numerous *safety* improvements, which include regularly updated flight/mission planning software, automated pre-flight checks that guarantee battery health, GPS reception, and level launch locations. The development of intelligent wind control systems allow algorithms to handle the wind dynamics, regardless of initial launch conditions. Many systems now include adaptive geofencing, which enables users to dynamically constrain the radius of an operation to prevent flyaways. Some of the more advanced user interfaces include real-time information on flight location, altitude, and mission progress, as well.

Lastly, although there is very little information publicly available regarding sUAS *reliability*, there are growing calls for these data. Interestingly, for military-grade systems, including the Predator and Reaper, there is some basic reliability information available (Petritoli et al. 2018). Further, there is some preliminary work directed toward isolating human-induced reliability problems, versus hardware-oriented issues with UAS (Tan et al. 2017). Unfortunately, there are no reliability data available on consumer or professional-grade sUAS. That said, Katz (2018) makes an important point regarding sUAS reliability for mapping and surveying, a point that can be easily extended to other domains. The primary concern of end-users should not be an sUAS that comes equipped with redundant onboard systems that help avoid a "single point of failure." Rather, end-users should purchase an sUAS that is so reliable, these redundant systems are not needed. Katz (2018) notes that mass-produced sUAS like the DJI Phantom have so many users that DJI's engineers get the opportunity to evaluate and troubleshoot the 1 in 10,000 reliability issue(s) that high-end manufacturers are never alerted to. In short, the high-end manufacturers do not have the user-base to do this, reducing their ability to estimate and trouble-shoot reliability problems. Further, Katz (2018) notes that because DJI produces closed systems, resistant to customization, but designed to work together, any modifications add unnecessary complexity, which can ultimately trickle down to field operations, data quality, and sUAS reliability.

1.3.1 What sUAS Cannot Do for Mapping and Urban Spatial Analysis

While sUAS are rapidly advancing mapping and spatial analysis, it is unrealistic to assume that they do everything well. As detailed by Casagrande et al. (2018), there are several notable limitations. First, and perhaps most important, sUAS have limited payload capabilities. This drastically limits the type and number of sensors that can be included on a mission. Although it would be amazing to mount a LiDAR system on a DJI Phantom 4 Pro, along with infrared, thermal, hyperspectral, and visible sensors, it simply is not possible, *yet*. Historically, LiDAR systems have been rather large, making it difficult to mount and operate with sUAS that have limited payload capacities. However, recent advances have made LiDAR drones a reality. For example, the Velodyne VLP-16 LiDAR puck weighs 830 g and provides 100 m range with a 360° horizontal field of view (FoV) and a 20° vertical FoV. Velodyne recently teamed up with Phoenix LiDAR Systems and mounted the VLP-16 to a DJI S1000 for real-time data collection (https://tinyurl.com/ybwoxcsm). That said, mounting all the desired sensors on a single, entry-level sUAS, at least for now, is not possible.

In addition to payload capacity limits, sUAS are limited in the amount of ground they can cover for a single flight. For example, under optimal conditions, the WingtraOne has a maximum flight time of 55 min and can cover approximately 400 ha when flying at 122 m altitude. The eBee X offers 90 min of flight time at 122 m altitude and can cover 500 ha. While this level of coverage is quite good for a single flight, there are obvious constraints. Depending on the size of a study area, multiple flights over multiple days may be required. This reduces the temporal consistency of the imagery and mapping outputs, but again, depending on the nature of the project, this may be acceptable.

There are also some limitations associated with the data generated by sUAS. While centimeter-level accuracy is sufficient for most cartographic and photogrammetric applications, there may be instances where sub-centimeter accuracy is required. For example, monitoring geologic deformations (Bellone et al., 2016) often requires sub-centimeter accuracy. This level of accuracy is not something that sUAS platforms are particularly good at doing, and that is ok. Although the WingtraOne gets close to centimeter-level accuracy in testing, it is important to remember that this level of accuracy is overkill for most mapping and spatial analysis applications.

Finally, the visible sensors used onboard sUAS that often struggle to generate viable digital terrain models (DTMs) using photogrammetric techniques. This is particularly true when the terrain surface is obfuscated by dense land cover (e.g., trees, vegetation). In these cases, LiDAR would be the most appropriate tool for use, but this can get expensive. The Velodyne VLP-16 LiDAR puck retails for $4000. Not an insurmountable cost for an sUAS sensor, but customizing a platform to accommodate the VLP-16 and store the generated data would require significant time, money, and effort.

1.4 Including an sUAS in a Technology Portfolio

Given the strengths and weaknesses associated with sUAS for mapping and spatial analysis, does it make sense to include one in a technology portfolio? This is a question that readers need to answer for themselves. But, there are some basic guidelines to keep in mind when making this decision. First, as detailed above, sUAS are especially good at collecting high-resolution data for small areas. If your study area is extremely large, satellite-generated data might be the best option. Second, if your study area requires sub-centimeter accuracy, alternative sensing options will be required. Third, sUAS are extremely good at collecting data for a study region with high-temporal frequency. If your study area requires repeated and temporally frequent measurements, there is no better platform than an sUAS. Fourth, because sUAS are extremely flexible and extensible, if researchers find themselves needing to work with multispectral and thermal sensors during the same day or week, the sUAS is a great option.

In addition to being concerned with the nature of sUAS measurement and the type(s) of outputs it can generate for a researcher, there are several more pragmatic concerns worth considering. For example, is the sUAS easy to fly? Crashing a $1500 piece of equipment is no fun. But, crashing a $30,000 piece of equipment is less fun. Flight planning software is not limited to the high-end sUAS, but if joysticks and manual controls are not something that make sense for a researcher, then perhaps a more significant initial investment will be necessary. This can be limiting. Further, who is going to fly the sUAS? As we detail in Chap. 2, there are several relatively significant hurdles in getting certified to fly an sUAS for research purposes in the United States. In addition to obtaining a Federal Aviation Administration's Part 107 certification, sUAS need to be registered, you need to use a trained spotter in the field (in addition to a pilot), and there are a variety of other constraints that limit *where*, *when*, and *how* one can fly an sUAS. In short, it is important to consider the overall time and energy required to qualify for flying an sUAS. As a researcher, do you have adequate computational equipment to process the sUAS outputs?

Generating point clouds that consist of a billion points is not trivial, nor is storing the generated data. In this way, sUAS often require secondary investments in computing and digital storage. Also worth considering is whether the sUAS manufacturer offers adequate customer support. If the drone is constantly offline because of hardware or software difficulties, its return on investment (ROI) is reduced. Speaking of ROI, blowing one's budget on the best sUAS one can afford never makes sense, especially if this means that no money is left for upkeep and routine maintenance. Further, will the sUAS advance one's research mission and mandate? If not, why bother?

In sum, sUAS are amazing technological platforms that can help facilitate mapping and spatial analysis in many ways. But, there is both *operational overhead* (e.g., cost, maintenance, certification) and *research overhead* (e.g., computing, data storage, software costs) associated with their use. However, if you have decided to purchase an sUAS for mapping and spatial analysis, the remainder of this book is for you.

1.5 Book Organization

The principle aim of this book is to provide readers with an introduction to the use of sUAS for spatial analysis and mapping. At its core, this book is intended to serve as a resource for educators and students who currently (or will in the future) work with sUAS, by addressing the knowledge, skills, and abilities required to efficiently use drones for obtaining high-quality data and generating actionable information. The velocity of urban change necessitates observation platforms that not only enhance situational awareness for planning and allied analytical efforts but also provide the ability to rapidly and inexpensively collect data and monitor change. sUAS can accomplish both tasks. However, their use in urban environments are loaded with social, operational, regulatory, and technical challenges that must be addressed for successful deployments. Again, this book provides a resource for researchers and students who work with geographic information and are seeking to enhance these data with the use of sUAS. The remainder of this book is organized as follows.

Chapter 2 provides an overview of sUAS platforms and discuss many different applications of sUAS in urban settings, from cargo delivery and transport to conducting comprehensive tree surveys in a neighborhood. Readers should consider this chapter a thorough, but not all-encompassing, literature review on sUAS for urban applications.

Chapter 3 provides a primer on reading aeronautical charts and maps. Interestingly, this is largely missing from the curricula in most geography departments around the USA. Although space limitations will prevent us from detailing all the nuances of these maps and charts, we provide readers with enough background to get them airborne, safely.

Chapter 4 details the legal, environmental, operational, and safety challenges in urban areas. Infinitely more complex than rural areas for flying, there is a huge range of restrictions and safety challenges that all pilots need to be aware of when operating in busy, urban skies.

Chapter 5 explores the sensitive issue of sUAS and personal privacy. As detailed by Nelson et al. (2019), privacy issues are one of the most significant concerns for the public when it comes to drones. We dig deeper into this issue, highlighting many nuances of personal privacy, policy, and the use of sUAS.

Chapter 6 focuses on operations and the process of integrating sUAS-derived data with geographic information systems (GIS). Issues associated with data formatting, portability, integration, computation, storage, and associated metadata protocols are detailed.

Chapter 7 provides the first of four case studies for the book, focusing on remotely sensing physical disorder in a neighborhood. Comparisons between more traditional satellite imagery and sUAS-derived imagery are used to highlight the advantages of drone use. Implications for urban planning are also addressed.

Chapter 8 is a case study that focuses on ways that unmanned aerial systems can help energy system maintenance and development. This includes a detailed case study of how one can use sUAS technology to conduct energy audits for residential neighborhoods. Ultimately, this information can help direct urban infrastructure investments to ameliorate dependence on fossil fuels and other highly pollutive energy sources.

Chapter 9 provides a case study on how sUAS can be used for detecting informal recreational trails. In addition, it highlights the ways in which unmanned aerial systems can contribute to efforts for identifying landscape fragmentation and potential interdiction points for minimizing the use of informal trails for enhancing landscape recovery.

Finally, Chap. 10 provides concluding thoughts and a discussion of urban futures with sUAS and other airborne sensing platforms. Applications for sUAS, policy implications, and issues of privacy are revisited, but with an eye toward future challenges.

References

Anderson, M. (2019). Precision agriculture from an eagle eye view. *Cavalier County Republican*. Retrieved October 13, 2019, from https://tinyurl.com/y8wqymcs

Bellone, T., Dabove, P., Manzino, A. M., & Taglioretti, C. (2016). Real-time monitoring for fast deformations using GNSS low-cost receivers. *Geomatics, Natural Hazards and Risk, 7*(2), 458–470.

Bishop, W., & Grubesic, T. H. (2016). *Geographic information*. Switzerland: Springer.

Casagrande, G., Sik, A., & Szabó, G. (Eds.). (2018). *Small Flying Drones*. Springer.

Ellenberg, A., Branco, L., Krick, A., Bartoli, I., & Kontsos, A. (2014). Use of unmanned aerial vehicle for quantitative infrastructure evaluation. *Journal of Infrastructure Systems, 21*(3), 04014054.

French, S. (2017). The gear you need for a drone mapping project is surprisingly cheap. *The Drone Girl*. Retrieved January 15, 2019, from https://tinyurl.com/y9m7rf5c

Giannini, C., Shaaban, A. A., Buratti, C., Verdone, R. (2016). Delay Tolerant Networking for smart city through drones. In *International Symposium on Wireless Communication Systems (ISWCS)*. IEEE.

Grubesic, T. H., Wallace, D., Chamberlain, A. W., & Nelson, J. R. (2018). Using unmanned aerial systems (UAS) for remotely sensing physical disorder in neighborhoods. *Landscape and Urban Planning, 169*, 148–159.

Goldman Sachs. (2018). *Drones reporting for work*. Retrieved September 18, 2018, from https://tinyurl.com/y7ztndv8

Katz, D. (2018). Surveyors and other professional drone operators should be focused on accuracy and reliability. *Commercial UAV News*. Retrieved June 20, 2018, from https://tinyurl.com/y7vkknn5

Kotha, S. (2010). Spillovers, spill-ins, and strategic entrepreneurship: America's first commercial jet airplane and Boeing's ascendancy in commercial aviation. *Strategic Entrepreneurship Journal, 4*(4), 284–306.

Levick, R. (2018). Drone industry just beginning to take off. *Forbes*. Retrieved July 8, 2019, from https://tinyurl.com/yc73kpff

Lipsey, R. G., Carlaw, K. I., & Bekar, C. T. (2005). *Economic transformations: General purpose technologies and long-term economic growth*. Oxford: OUP.

Lu, Y., Macias, D., Dean, Z. S., Kreger, N. R., & Wong, P. K. (2015). A UAV-mounted whole cell biosensor system for environmental monitoring applications. *IEEE Transactions on Nanobioscience, 14*, 811–817.

Manfreda, S., McCabe, M., Miller, P., Lucas, R., Pajuelo Madrigal, V., Mallinis, G., & Müllerová, J. (2018). On the use of unmanned aerial systems for environmental monitoring. *Remote Sensing, 10*(4), 641.

Maza, I., Kondak, K., Bernard, M., Ollero, A. (2009). Multi-UAV cooperation and control for load transportation and deployment. In *Selected papers from the 2nd International Symposium on UAVs, Reno, Nevada, USA June 8–10, 2009, Springer, Dordrecht.*

Merriam-Webster Dictionary [MW]. 2019. *Drone*. Retrieved from https://tinyurl.com/y95nvmv8

Murphy, D. W., Cycon, J. (1999). Applications for mini VTOL UAV for law enforcement. In *Sensors, C3I, information, and training technologies for law enforcement*. (Vol. 3577, pp. 35–44). Bellingham: International Society for Optics and Photonics.

Nelson, J. R., Grubesic, T. H., Wallace, D., & Chamberlain, A. W. (2019). The view from above: A survey of the public's perception of unmanned aerial vehicles and privacy. *Journal of Urban Technology, 26*(1), 83–105.

Nex, F., & Remondino, F. (2014). UAV for 3D mapping applications: A review. *Applied Geomatics, 6*(1), 1–15.

Niosi, J., & Zhegu, M. (2010). Multinational corporations, value chains and knowledge spillovers in the global aircraft industry. *Institutions and Economies, 2*, 109–141.

Orfanus, D., de Freitas, E. P. (2014). Comparison of UAV-based reconnaissance systems performance using realistic mobility models. In *6th International Congress on Ultra Modern Telecommunications and Control Systems and Workshops (ICUMT)*, IEEE (pp. 248–253).

Petritoli, E., Leccese, F., & Ciani, L. (2018). Reliability and maintenance analysis of unmanned aerial vehicles. *Sensors, 18*(9), 3171.

Rokhmana, C. A. (2015). The potential of UAV-based remote sensing for supporting precision agriculture in Indonesia. *Procedia Environmental Sciences, 24*, 245–253.

Robinson, A. H., Morrison, J. L., Muehrcke, P. C., Kimerling, A. J., & Guptill, S. C. (2005). *Elements of Cartography*–6th Ed, 544 pp. New York, John Willey & Sons.

Schenkelberg, F. (2016, January). How reliable does a delivery drone have to be. In *2016 Annual Reliability and Maintainability Symposium (RAMS)* (p. 978). IEEE.

Silvagni, M., Tonoli, A., Zenerino, E., & Chiaberge, M. (2017). Multipurpose UAV for search and rescue operations in mountain avalanche events. *Geomatics, Natural Hazards and Risk, 8*(1), 18–33.

Sydell, L. (2017). Telecom companies turn to drones for help after hurricanes. *NPR*. Retrieved December 12, 2018, from https://tinyurl.com/y9zlvxzw

Tan, Y., Feng, D., Shen, H. (2017). Research for unmanned aerial vehicle components reliability evaluation model considering the influences of human factors. In *MATEC Web of Conferences* (Vol. 139, p. 00221). EDP Sciences.

Tripicchio, P., Satler, M., Dabisias, G., Ruffaldi, E., Avizzano, C. A. (2015). Towards smart farming and sustainable agriculture with drones. In *International Conference on Intelligent Environments (IE)* (pp. 140–143). IEEE.

Webb, P., Mehlhorn, S. A., Smartt, P. (2017). Developing protocols for using a UAV to monitor herd health. In *ASABE Annual International Meeting* (p. 1). American Society of Agricultural and Biological Engineers.

WIPO [World Intellectual Property Organization]. (2015). *Breakthrough innovation and economic growth*. Retrieved from https://tinyurl.com/ol42v9u

Zacharek, S. (2018 May 18). How drones are revolutionizing the way film and television is made. *Time*. Retrieved from https://tinyurl.com/ycskbvgb

Chapter 2
UAS Platforms and Applications for Mapping and Urban Analysis

Abstract The use of unmanned aerial systems (sUAS) is growing rapidly. sUAS are used in a wide range of applications, including emergency response, law enforcement, infrastructure inspection and management, as well as mapping and surveying. In this chapter, we provide an overview of sUAS platforms and discuss many different applications of sUAS in urban (or near urban) settings, from public health to cargo transport and delivery. Readers should consider this chapter a thorough, but not all-encompassing, literature review on sUAS for urban applications.

Keywords Drone · Remotely piloted vehicles · Fixed wing · Rotary wing · Multi-rotor · Green infrastructure · Environmental monitoring · Public safety · Public health · Infrastructure inspection · Law enforcement · Transport · Mapping

2.1 Introduction

Small unmanned aerial systems (sUAS), unmanned aerial vehicles (UAVs), remotely piloted vehicles (RPVs), and drones are terms that are often used interchangeably about aircraft that are operated without direct human intervention *within* the aircraft itself. This is unfortunate, as there are several subtle (and not so subtle) differences between these platforms. The most widely adopted term for these vehicles, especially in the media, is "drone." As noted by Fahlstrom and Gleason (2012, p. 7), "drone" is a term that is not used in the UAV development and user communities unless it is referring to a vehicle that has limited flexibility for accomplishing a mission and flies in a "persistently dull, monotonous and indifferent manner." Military target drones are a good example of this. For example, Rees (2018) outlines the specifications of the M-40 target drone, developed by Leonardo, which is used by the Italian Directorate of Air Armaments and Airworthiness. In short, the M-40 is designed to mimic a variety of aircraft and missiles. It includes onboard radar and infrared systems, can perform low-altitude sea skimming missions and high-G maneuvers during its 60 min of flight endurance. While the onboard components are sophisticated, and its ability to pull high-G maneuvers is certainly not dull, the drone has a single purpose—serving as a target.

T. H. Grubesic, J. R. Nelson, *UAVs and Urban Spatial Analysis*,
https://doi.org/10.1007/978-3-030-35865-5_2

Conversely, RPVs are actively controlled by humans from a distance. The Predator XP (http://www.ga-asi.com/predator-xp) by General Atomics Aeronautical System Incorporated is a good example of a RPV. In addition to having weapons onboard, the Predator XP includes multiple intelligence, surveillance, and reconnaissance (ISR) sensors that include multi-mode radar and electro-optical infrared cameras (Fuller 2017). Ultimately, pilots are charged with not only flying the vehicle and monitoring the sensor systems but also deploying the weapons, when necessary. To put this in context, Fahlstrom and Gleason (2012) note that RPVs are always considered UAVs, but since UAVs can perform autonomous or preprogrammed missions, they are not always RPVs.

For the purposes of this book, we focus on small, unmanned aerial systems (sUAS). This is important for several reasons. First, as detailed by Terwilliger et al. (2017), UAVs are increasingly recognized as the *aerial* element of a unified system. However, unmanned aerial system (UAS) better describes the *system of systems* onboard the UAS. As detailed in Chap. 1, this includes the aerial platform itself, the payload, control and communication systems, support equipment, and the proprioceptive and exteroceptive sensors. Second, *small* UAS are generally defined as platforms that weight less than 55 pounds (~25 kg), with all systems accounted for. Not only are these small systems affordable, they are highly customizable. As a result, sUAS are not strictly limited to hobbyists and recreational flying. Today, sUAS include a wide range of consumer- and professional-grade options that can be used in a variety of applications, including precision agriculture, crop treatment, livestock monitoring, emergency response, law enforcement, infrastructure inspection and management, as well as mapping and surveying. Finally, it is important to note that we will use the terms sUAS, UAS, UAV, and drone in a mixed fashion throughout the book. We understand the differences (and readers should too), but there are only so many times that one can put sUAS in a sentence or paragraph.

The purpose of this chapter is to provide a broad overview of sUAS applications and their use in urban spatial analysis. We begin by briefly detailing the different types of sUAS platforms and providing readers with information on how these platforms can be used/applied across a variety of domains. This is followed by a more detailed review of sUAS applications in urban settings. Again, readers should consider this chapter a thorough and practical review of the literature on urban applications for sUAS, but it is not all-encompassing.

2.2 sUAS Platforms

There are three basic sUAS platforms available in the commercial market: (1) *fixed wing*, (2) *rotary wing*, and (3) *multi-rotor*. Fixed wing systems are configured like an aircraft with some type of propeller to produce thrust. The wings are rigid and have a predetermined airfoil. There are numerous advantages to

Fig. 2.1 eBee classic fixed wing drone

fixed-wing systems, including power efficiency, the ability to glide, longer flight endurance, and the ability to cover larger areas in a single flight. Its simple structure also provides for a less complicated maintenance and repair process, often increasing operational time in the field. The aerodynamics of fixed wings also means they are fast. For example, Fig. 2.1 displays eBee Classic fixed wing sUAS from Sensefly, which is a typical fixed wing configuration that uses a rear propeller. The eBee weighs ~1.5 pounds (700 g), and cruises at speeds between 25 and 56 mph (40–90 km/h). Under normal operating conditions, the eBee can cover approximately 500 ha in a single flight. Later in the book, we will highlight some results generated by an eBee Classic, so that readers can get a better idea of the quality of the data generated by this fixed wing sUAS. The major disadvantage of using most fixed-wing sUAS is that they must be launched in an area with plenty of open space, often with some type of launching mechanism. Fixed wings also require an area that can accommodate a soft, belly landing at the conclusion of the flight. Recent advances in vertical takeoff/landing (VTOL) sUAS, such as the WingtraOne (https://wingtra.com/drone/), have made progress in ameliorating these constraints.

Rotary wing sUAS utilize a single rotor to generate thrust. There are several advantages to this, including vertical takeoff/landing and the ability to carry heavier payloads, which is ideal for package delivery and mounting heavier sensors (e.g., light detection and ranging, LiDAR) onboard. Rotary wing sUAS also allow for longer flight times than multi-rotor platforms, and an ability to fly considerably faster than multi-rotor drones along a linear path. For example, consider the Prodrone PDH-01, a long-distance delivery platform (Fig. 2.2). It weighs 19.84 lbs (9 kg) and has a propeller diameter of 70 inches, but it can fly at speeds over 90 mph (144 km/h). Its specifications also indicate that it has a flight duration of 60 min and the ability to carry a payload of 6.61 lbs (3 kg) (Matsunaga 2017). Yes,

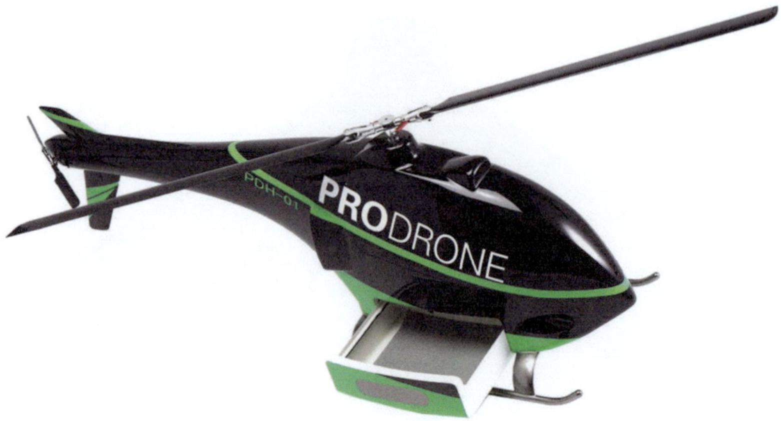

Fig. 2.2 Prodrone PDH-01 rotary delivery drone

the PDH-01 is gasoline powered (somewhat atypical for sUAS) and not nearly as fast as the Prodrone PDH-03 (130 mph), but its ability to carry nearly 3 kg of payload is significant.

Finally, multi-rotor sUAS are available in a wide range of configurations and consist of multiple rotors for generating lift. For example, while platforms using a single rotor (e.g., Prodrone PDH-01) are referred to as a helicopter, platforms that use three rotors are known as tricopters, while four, six, and eight rotors are known as quadcopters, hexacopters, and octocopters, respectively. There are numerous alternatives (e.g., 12 or 16 rotors), as well as Y6 configurations, which is a tricopter that has twin rotors on each arm, one facing up and the other down. The major advantages to using multi-rotor sUAS are threefold. First, they allow for vertical takeoff/landing for operations in space-constrained launch sites, including boat decks and dense urban areas. Second, they are highly stable in flight. As a result, multi-rotors are the platform of choice for professional cinematography and real estate photography. This stability also provides an advantage when conducting visual inspections of infrastructure such as radio towers and bridges. Third, they can carry extremely large payloads. For example, consider the DJI Agras MG-1 octocopter, which is capable of carrying up to 22 lbs of liquid fertilizer or pesticides for distribution in agricultural settings (Fig. 2.3). DJI claims that Agras MG-1 is 40 times faster than manual spraying and can cover 4000–6000 sq m in less than 10 min (Byford 2015). The major disadvantages of multi-rotors include more moving parts and more maintenance/upkeep. In addition, they are much slower than single-rotor or fixed wing sUAS, which translates to more flights and longer-duration operational windows to cover larger areas.

Fig. 2.3 DJI Agras MG-1 octocopter

2.3 Applications of sUAS in Urban Settings

Terwilliger et al. (2017) outline three key aspects of an effective sUAS application that must be considered early in the research development process: (1) *satisfies needs*, (2) *provides benefit*, and (3) *improvement*. The first consideration for urban applications must always be whether the sUAS *satisfies* needs. Specifically, does it provide a feasible solution to solving a real problem? As detailed in Chap. 1, economic considerations are critical here, because many of the most advanced and efficient sUAS platforms require a significant initial investment. For example, in early 2019, the DJI Phantom 4 Pro (2.0 quadcopter) (https://tinyurl.com/y7ngsg34), which includes a 20 megapixel CMOS sensor and obstacle detection, costs US$1499. While the entry cost of this model is relatively low, its flight times are limited (30 min), and its onboard sensor is of middling quality. Conversely, the Altavian Galaxy R8700 (https://tinyurl.com/y8sdg68z) starts at US$15,000, but can include a variety of sensors. The base package will provide a Canon SL1 camera and NovAtel OEMStar GPS which achieves 0.7 m resolution for mapping applications. More advanced sensor packages for the Altavian can provide 1 cm resolution, post-processing kinetics, and the ability to collect color-infrared (CIR) data. The questions at hand is a simple one—how much drone do you need? If the smaller and less expensive Phantom 4 Pro can do the job, it may be the best choice (Katz 2018).

The second consideration concerns the provided *benefits* of using an sUAS. Specifically, does the conceptual solution that requires an sUAS establish more advantages than it does limitations or problems? For example, if a city is interested in evaluating the physical deterioration of sidewalks in several neighborhoods, does the use of a drone provide any tangible benefits versus visual inspection and documentation by city workers? In this instance, the advantages of an sUAS include the surveying of sidewalk areas from an elevated perspective, large area coverage with minimal time (e.g., 30 min) commitments, and an improvement in precision where repetitious missions are required. The disadvantages include the presence of seasonal tree cover (obscuring sidewalks), the cost to purchase an sUAS and have a trained pilot and spotter on staff, and the need to obtain flight permissions over areas exhibiting dense pedestrian flows or in tightly regulated airspace.

Third, would the use of an sUAS result in a quantifiable *improvement* in operations or geospatial intelligence when compared to the current methods? Again, issues of cost are relevant here, as are issues of risk. For example, the process of conducting infrastructure integrity evaluations, especially for large radio towers, electrical transmission lines, wind turbines, and bridges are especially risky for human personnel because of the risks associated with falls, electric shock, and related hazards. If the sUAS can reduce these risks, improving data generation and associated outcomes, it might lead to improvement over current methods.

A fourth aspect, not detailed by Terwilliger et al. (2017), but key to urban operations, is evaluating the level of *convenience* in flying an sUAS. This is especially true in locales (e.g., countries, states, cities) where sUAS operations are regulated. Further, as will be discussed in Chap. 4, the sheer number of operational considerations for flying in urban areas may create a critical mass of problems so significant, non-sUAS solutions are more appealing. For example, pilots and spotters must maintain a visible line of sight (VLOS) to the sUAS (at all times), which can be difficult to achieve in dense urban areas filled with buildings and other types of infrastructure that can obfuscate VLOS. The sUAS can only be operated at 400 ft above ground level, must yield right-of-way to all other aircraft, and cannot be flown at night. It is possible to get a waiver for some of these things, but that also requires effort.

Given this wide range of considerations, there are many applications for sUAS in urban areas. In the following subsections, we detail several of the most prevalent uses for sUAS in mapping and urban analysis.

2.3.1 Green Infrastructure

Green infrastructure is a term that describes how green spaces function together for providing ecological services to a neighborhood, city or larger region. Although there are a variety of definitions for green infrastructure found throughout the literature, all definitions generally look something like this (Natural England 2010, p. 8):

> A strategically planned and delivered network comprising the broadest range of high quality green spaces and other environmental features. Designed and managed as a multifunctional resource capable of delivering those ecological services and quality-of-life benefits required by the communities it serves and needed to underpin sustainability. Its design and management should also respect and enhance the character and distinctiveness of an area with regard to habitats and landscape types.

Recent work that uses sUAS for the evaluation of green infrastructure is quite varied. For example, Gadi et al. (2017) use imagery collected by an sUAS to help evaluate the spatial and temporal variations in hydraulic conductivity for vegetated soil (i.e., mixed grass cover) near trees. The sUAS (a DJI Phantom) was specifically used to assess vegetation density. In another work, Hassaan et al. (2016) use a DJI

Phantom, generated RGB mosaics, and image segmentation to count the number of tree issues in an area proximal to Lahore, Pakistan. Liang et al. (2017) implement a three-dimensional green quantity index to evaluate the extent and the environmental and climactic benefits of urban green space. The authors used a fixed-wing, gasoline-fired sUAS to collect 426 high-resolution images. Additional research on green infrastructure which emphasizes treescapes includes the work of Lin et al. (2015) who developed a new method that combines *k*-means clustering, RGB-based vegetation index derivation, and new feature parameters for synthesizing texture and color for delineating trees and marker-control techniques for evaluating canopy shapes and detecting individual trees in oblique sUAS-derived imagery. Finally, Gaitani et al. (2017) use a combination of multispectral and infrared sensors on an sUAS to identify reflectance properties of various urban settings throughout Athens, Greece. The authors also explore the response of vegetation to evaluate the climate parameters essential to bio-climactic design aims.

2.3.2 Environmental Monitoring

Similar to efforts directed toward evaluating green infrastructure, there is a growing literature pertaining to environmental monitoring with sUAS. Much of this work occurs outside of urban locales. Manfreda et al. (2018) provide the most current review on the topic, emphasizing natural and agricultural ecosystem monitoring. In a more focused topical review, Bhardwaj et al. (2016) detail sUAS use in glaciology, noting many advantages associated with cost to deploy, quality of derived data, and the flexibility/extensibility of sUAS sensors. A similar literature exists in evaluating and managing rivers, where sUAS are used to monitor channel displacement and bank erosion (Hemmelder et al. 2018). More urban-focused applications include the use of sUAS to monitor tree health. For example, Näsi et al. (2018) develop ways to assess and compare bark beetle infestations at the individual tree level in urban forests. The authors use a hyperspectral camera mounted on an sUAS and compared the results against hyperspectral sensors mounted on a Cessna airplane. Ultimately, the finer-resolution sUAS dataset provided better results for estimating tree health.

Park and Um (2018) use both visible and near infrared (NIR) sensors mounted on an sUAS to identify the major types of carbon sinks versus sources in/around a university campus in an urban area. In some very early simulation work, Sinha et al. (2009) propose the development and use of a mechanism to detect, model, and track the shape of a contaminant cloud boundary using sensors mounted to sUAS. Finally, Rossi et al. (2014) recommend the use of sUAS to measure spatially distributed gas concentrations, with an eye toward tracking volatile chemical concentrations. They suggest that their proposed gas sensing system would be fully autonomous and allow for simultaneous localization and mapping (SLAM) of the volatiles. For readers unfamiliar with SLAM, it refers to the problem of constructing and updating a map in an unknown environment while simultaneously tracking the location of an agent, within that unknown environment (Durrant-Whyte and Bailey 2006; Bailey and Durrant-Whyte 2006).

2.3.3 *Public Safety and Emergency Response*

One of the most vibrant areas of sUAS research for urban areas exists within the domain of public safety and emergency response. This occurs in many different substantive fields, from transportation to disaster science, but the common thread for all of this work is the innovative use of sUAS to improve outcomes. Consider, for example, the contribution of Gu et al. (2019) which utilizes sUAS video data for evaluating automobile crash risks in freeway interchange merging areas. For this application, the authors used a DJI Inspire 1, collecting aerial video 80 m upstream and 100 m downstream of the acceleration lane. Their results suggest that sUAS is an excellent tool for this type of work. Not only can the sUAS continuously fly at a proper height, it is able to cover a large area, flexibly, with minimal cost and without affecting drivers' behavior. Not surprisingly, there is a significant amount of work in traffic-based public safety analysis using sUAS (Kanistras et al. 2015; Khan et al. 2017, 2018; Salvo et al. 2014).

Unmanned aerial systems are also being applied widely in emergency response and disaster recovery. Consider the work of de Oliveira Silva et al. (2017), which details the benefits of UAS use for speeding up the information collection process post-disaster. The combination of UAS imagery and geographic information systems (GIS) was used to develop a procedure to support the design of aid distribution networks in the municipality of Duque de Caxias, Brazil, including the identification of locations for points of service. Rabta et al. (2018) outline many ways in which sUAS are receiving increased attention from humanitarian organizations for last-mile distribution problems. In short, getting the needed medical supplies, food, and water can prove difficult after disasters because some neighborhoods are cutoff from main roads. Rabta et al. (2018) formulate an optimization model for delivering multiple packages of lightweight relief items, including vaccines and water purifications tablets that minimize the total travel distance of the sUAS. Important constraints including payloads, energy (i.e., battery life for flight duration), and the placement of recharging stations are also considered. Tuna et al. (2014) propose the use of sUAS for reestablishing communication systems for essential rescue operations, post-disaster. The developed system includes an end-to-end communication system, an sUAS navigation system, and a formation control strategy for the airborne sUAS. Simulation results and testing with a UAV suggest that the system is promising, but requires more testing to account for jitter, throughput, availability range, and packet losses. For applications in urban areas, the system would also need to be tested for interference caused by reflective structures and buildings.

Finally, in the state of Oregon, sUAS were used in a search and rescue mission in the Columbia Gorge, outside of Portland, Oregon (Van Tilburg 2017). Specifically, in a joint effort, the Cascade Locks Fire Department and the Crag Rats Mountain Rescue team were trying to determine if a man who had slipped into a slot canyon during a recreational climb was injured or worse. No vocal contact could be made with the victim. Efforts to rappel into the canyon were halted because the requisite safety equipment was not available to the emergency response teams. The team decided to use a DJI drone for obtaining visual contact with the victim, ultimately confirming a fatality. The body was recovered the next day by mountain rescue crews (Van Tilburg 2017). In this instance, the ability to leverage sUAS technology prevented first responders from making a dangerous descent into the slot canyon, seriously risking life, and limb without the proper equipment, when determining the severity of the accident.

2.3.4 Public Health

The diversity of sUAS applications in public health is growing. As might expect, there is some modest overlap between public health and emergency response applications. For example, as detailed above, the work of Rabta et al. (2018) and their optimization model for delivering lightweight relief items (e.g., vaccines, water purification tablets) is important for maintaining public health in disaster situations. Haidari et al. (2016) also explore the utility of sUAS for delivering vaccines in the developing world. The authors note that many of the advantages typically associated with sUAS, including their ability to traverse difficult terrain, reduce labor, and replace fleets of maintenance intensive vehicles, are important considerations when evaluating their use for vaccine delivery. Using modeling data in the Gaza province, their results suggest that sUAS deliveries not only has the potential to increase vaccine availability, but if the drone infrastructure is used regularly, the initial capital costs of deployment can be recovered and cost savings of ~$0.05 and $0.21 per dose can be realized. In Rwanda, the government recently announced emergency drone delivery service for carrying blood supplies to clinics in need (Rutkin 2016). Program directors note that some areas of Rwanda are extremely difficult to access, especially during the rainy season, where the use of roads is simply not an option. With the integration of sUAS technology, when health workers at any of the 21 national clinics need a delivery of blood, they simply text the drone base, and an sUAS is loaded with up to 1.5 kg of blood for a parachute drop delivery (Rutkin 2016).

Another application that straddles both emergency response and public health pertains to the geographic positioning of medical drones in urban areas (Pulver et al. 2016). In recent work, Pulver and Wei (2018) detail a flexible, spatial optimization model for locating medical drones with onboard, automated external defibrillators (AED). Their results indicate that the use of 71 drones with 68 launch sites can cover 90.4% of the historical, out-of-hospital cardiac arrests in Salt Lake City. But more importantly, the response time for the sited drones is constrained to 1 min or less—an important consideration when considering survival rates for cardiac arrest (Larsen et al. 1993). Related work on the use of sUAS for delivering AEDs can be found in Claesson et al. (2016)

In a recent simulation study, Seguin et al. (2018) evaluate the efficiency of sUAS for providing floatation devices to potential drowning victims. The results suggest that the sUAS drop strategy significantly reduces the time for a drowning victim to access a floatation device when compared to both standard and jet ski rescue operations by lifeguards. More importantly, the sUAS option was more effective in calm, moderate, and rough sea conditions. Claesson et al. (2017) also explore the use of sUAS for helping to locate drowning victims, with a specific goal of preventing out of hospital cardiac arrest.

2.3.5 *Infrastructure Inspection*

Perhaps the premier application for sUAS in urban areas is their use for infrastructure inspection. For example, Inzerillo et al. (2018) utilize sUAS and structure from motion (SfM) to both detect and analyze road pavement distress. Although the sUAS-derived information was not as finely tuned or as accurate as LiDAR, the authors suggest that the sUAS allows for a much faster identification of pavement distress and the overall condition of the road, on a large scale. Pavement conditions, however, are just one facet of a diverse infrastructure ecosystem that includes buildings, the electrical grid, telecommunications, dams, bridges, and the like. Central to these infrastructure elements is the construction industry—charged with both building and maintaining many of these systems. Tatum and Liu (2017) document the growth of sUAS in the construction industry. Their survey results suggest that many US companies have integrated drones into their operation, using sUAS-derived imagery to replace traditional still photo and video acquisition. More importantly, much of this imagery is being processed with photogrammetric software and integrated into advanced building information models. For an excellent review on the applications of sUAS for the built environment, with an eye toward automated building inspections, see Rakha and Gorodetsky (2018).

Interestingly, the literature on infrastructure inspection largely falls outside of academic journals. However, there are thousands of examples available in the gray literature. For instance, the Black & Veatch engineering firm recently announced the successful conclusion of a 60-mile long, non-stop, infrastructure evaluation of electrical transmission lines (Sawyer 2018). The most interesting aspect of this effort was that the inspection included operations beyond visual line of sight (BVLOS). In the petroleum industry, Shell Global recently announced plans for deploying sUAS to inspect the equipment at the 200-acre (44 building) Shell Technology Center in Houston. Using Kesprey sUAS, the goal is to inspect a wide variety of equipment, faster, including equipment that may be too dangerous for employees to be near (Rattigan 2019). Intel, best known for making computer processors, recently inked a deal with the Minnesota Department of Transportation and the Kentucky Transportation Cabinet to use sUAS for bridge inspections in both states (Dormehl, 2018). Intel's Falcon 8+ (https://tinyurl.com/yafbnabh) platform will be used. Additional infrastructure inspection efforts include thermal imaging/inspections for buildings (Margaritoff 2018) and the use of drones to both audit and inspect telecommunication towers—which is widely recognized as one of the most dangerous jobs in the industry (Tomas 2017).

2.3.6 Law Enforcement

The use of sUAS for law enforcement overlaps many of the areas already covered, from traffic observation, to emergency response and search and rescue. For example, consider the somewhat controversial deployment of drones by the Los Angeles Police Department (LAPD). In a recent confrontation in Koreatown, which lasted 9 h, the LAPD used a DJI Spark to collect real-time information on the whereabouts of a suspect in a second story apartment (Chang 2019). LAPD anticipates that sUAS will prove especially helpful for hostage situations, response to natural disasters, and hazardous material incidents, as well as searches for suspects that may have "superior firepower" or an "extraordinary tactical advantage" (Chang 2019). It is also important to note that the use of drones by the LAPD has been controversial in the community, with residents fearing mission creep and the loss of privacy. This is exactly what Anania et al. (2019) address in their recent article, which explores levels of public support for police drone missions. The results suggest that study participants were more supportive of police use of drones when flying over predominantly African-American neighborhoods, when compared to primarily Caucasian neighborhoods. Further, participants identified as political liberals were less supportive of law enforcement sUAS use when compared to participants who were more politically conservative (Anania et al. 2019).

In other applications of sUAS for law enforcement, the Polk County Sherriff's Office (Florida) has developed an Aerial Response Team (ART), consisting of DJI Mavic drones that are deployed during a variety of emergency situations, including efforts to find missing persons and tracking suspects who fled. In the year since ART has been operating, the agency estimates a savings of nearly $107,000 in operating costs by keeping helicopters grounded and replacing them with sUAS for operations (Berkowitz 2018). Lastly, there is some work emerging on the use of traffic *enforcement* drones (Rosenfeld 2019). To be clear, rather than the standard application of using drones to monitor traffic (Kanistras et al. 2015; Khan et al. 2017, 2018; Salvo et al. 2014), enforcement drones are tasked to enforce local rules of the road. For example, Canada's Ministry of Transportation is collaborating with the private sector and the University of Toronto to build an sUAS monitoring framework that will help monitor and enforce the use of high-occupancy vehicle/high-occupancy toll lanes. In survey work conducted by Rosenfeld (2019) regarding traffic enforcement drones, the results suggest that respondents would be more comfortable with this application if they were used, at least initially, in interurban spaces, adhered to a privacy preserving policy and were used to enforce widely approved offences (e.g., aggressive driving, red light running, and identifying stolen vehicles).

2.3.7 *Cargo Transport and Delivery*

Much of the recent works regarding the UAS and cargo transport/deliver focus on strategic approaches for improving the efficiencies (e.g., geographic, operational, costs) associated with transport efforts. In this context, there are several facets of sUAS applications for cargo delivery and transport that must be considered. First, it is important for readers to remember that UAS load carrying capacities, at least in 2019, remain quite low. While most can handle several kilograms, the best-in-class multi-rotors, such as the Versadrones Heavy Lift Octocopter (https://tinyurl.com/y852596u), has a payload capacity of 12 kg. To put this in perspective, A Ford Transit 350 cargo van has a payload of approximately 2040 kg, 170 times greater than the Versadrone UAS. More importantly, unlike drones, a fleet of cargo vans can serve an entire city from a single location and are rarely affected by rain, snow, or inclement weather. That said, even with these limitations, UAS cargo operations have their own advantages, including rapid, deployment, low fleet, and maintenance costs, as well as imperviousness to surface transport delays (i.e., traffic jams).

Given these upsides to UAS cargo delivery and transport, but with an understanding of the known constraints, Chauhan et al. (2019) developed an approach to maximize the geographic coverage of drones, while simultaneously accounting for capacity-constrained facilities and range-constrained drones. Although Gurobi, a state-of-the-art mixed integer programming (MIP) solver, was able to identify extremely high-quality solutions for drone placements, the long running times limit real-world applicability. To mitigate this limitation, Chauhan et al. (2019) developed a three-stage heuristic that obtained solutions within 5% of the best Gurobi solutions, but required significantly less computational effort.

Chang and Lee (2018) developed a mathematical programming approach to optimize delivery routing of trucks that serve as a mobile, home-base for a handful of sUAS that will make cargo deliveries along the truck's route. This amounts to a multi-modal (truck + drone) mobile delivery depot. Related work includes that of Murray and Chu (2015), who explore the traveling salesman problem to optimize the routing for standard delivery vehicles, but one that can also include a "flying sidekick" (sUAS) for drone-assisted parcel delivery.

Hong et al. (2018) develop an optimization model for locating sUAS recharging stations for a commercial drone delivery service, integrating the Euclidean shortest path routing to ensure that sUAS avoids obstacles in the region it is serving. Faust et al. (2017) present a learning approach to help cargo-bearing sUAS deal with static obstacles in the environment and maintain a collision-free path through space.

Finally, Goodchild and Toy (2018) examine the differences in CO_2 emission and vehicle-miles traveled (VMT) for two delivery models, the first being trucks and the second being sUAS. The results were mixed and largely contingent upon the energy requirements of the drone. There were a handful of key takeaways. First, sUAS generally have a CO_2 emissions advantage over trucks, but this is for service zones that are closer to the drone depot, or have a fewer number of recipients. That said, trucks have a CO_2 emission advantage when service zones are far away or have a high number of recipients. Goodchild and Toy (2018) also note that sUAS significantly outpaced trucks for VMT because they were forced to return to the depot after every stop. In sum, a blended system that incorporates both sUAS and trucks would function best, with sUAS handling shorter distance deliveries and trucks handling those further away from the depot.

2.3.8 *Mapping and Surveying*

In one way or another, virtually every application detailed in this chapter involves some form of mapping, either directly or indirectly. For example, all the spatial optimization models detailed in the previous section must account for geographic space in their calculations. The same can be said for efforts to evaluate green infrastructure, conduct environmental monitoring, and efforts directed at search and rescue. For applications dealing with infrastructure inspection, although the geographic nuances of the infrastructure location may not be directly relevant, the sUAS must navigate in/around the infrastructure under observation, which of course requires mapping. This is not to say that direct applications of mapping and surveying are sparse in the literature, quite the opposite is true. Nex and Remondino (2014) provide a comprehensive (but quickly becoming outdated) review of the use of sUAS for 3D mapping applications. In more substantively focused work, Feng et al. (2015) leverage the image collection capabilities of an sUAS to conduct urban flood mapping in heterogeneous landscapes in Yuyao City, China. The results show that the inclusion of texture features can significantly improve the classification accuracy of the mapped, flooded locations. Gevaert et al. (2017) use sUAS-generated point clouds and 3D geometric features derived from sUAS imagery for identifying informal settlements in Kigali, Rwanda and Maldonado, Uruguay. In particular, the dense 3D point clouds and a high-resolution orthomosaic were exploited to better understand the local scene.

Similar to the field of infrastructure inspection, much of the innovation and application of sUAS for mapping and surveying occur in the private sector and can be found in the gray literature. For example, Karpowicz (2018) details efforts to accurately and consistently stereo compile from imagery collected by inexpensive UAS systems. The core benefits of stereo compilation are many, but the most alluring benefit is associated with photo-interpretation and mapping. Humans are notoriously bad at interpreting two-dimensional images (routinely misreading scale on both maps and imagery), but they are much better at interpreting 3D imagery. The challenge is to derive imagery that is of consistently high quality to allow for stereo compilation. In the developing world, sUAS are being used to map previously unmapped areas, including portions of Tanzania and the high growth areas in/around Zanzibar (Dwyer 2019). In Calgary, Alberta, a project to support the development of a new graveyard site was completed using an eBee Plus for mapping, but also included BVLOS operations (Lillian 2018).

2.4 Conclusion

This chapter covered a lot of ground, attempting to highlight the many different applications of sUAS in urban areas. Here is a scary thought for readers—we have only scratched the surface with this review. We completely ignored agricultural applications (e.g., pollen delivery), efforts to track polar ice sheets, monitor and observe physical activity in parks, monitor and predict the spread of wildfires, monitor, protect, and manage livestock, as well as the use of drones to smuggle illegal substances. For obvious reasons, we did not cover the emergence of unmanned underwater vehicles or many earth-bound drone technologies used to help inspect underground infrastructure systems, such as sewers. Limiting our scope was both intentional and necessary for this book. But, readers should not feel constrained by our limitations. We encourage you to explore the related literature at your leisure, knowing that much of it will have strong ties and interrelationships to the work covered here. Further, with each of these domains growing quickly, there is little doubt that the review provided here will be soon outdated.

References

Anania, E. C., Rice, S., Pierce, M., Winter, S. R., Capps, J., Walters, N. W., & Milner, M. N. (2019). Public support for police drone missions depends on political affiliation and neighborhood demographics. *Technology in Society, 57*, 95–103.

Bailey, T., & Durrant-Whyte, H. (2006). Simultaneous localization and mapping (SLAM): Part II. *IEEE Robotics & Automation Magazine, 13*(3), 108–117.

Berkowitz, K. L. (2018). Law enforcement: Use of drones saves time, money. *The Ledger*. Retrieved September 23, 2019, from https://tinyurl.com/y8aoqeb4

Bhardwaj, A., Sam, L., Martín-Torres, F. J., & Kumar, R. (2016). UAVs as remote sensing platform in glaciology: Present applications and future prospects. *Remote Sensing of Environment, 175*, 196–204.

Byford, S. (2015). DJI announces $15,000 agricultural drone designed to spray crops. *The Verge*. Retrieved February 15, 2018, from https://tinyurl.com/y78fdmjd

Chang, C. (2019). LAPD deploys controversial drone for the first time. *Los Angeles Times*. Retrieved from https://tinyurl.com/y83oq9o3

Chang, Y. S., & Lee, H. J. (2018). Optimal delivery routing with wider drone-delivery areas along a shorter truck-route. *Expert Systems with Applications, 104*, 307–317.

Chauhan, D., Unnikrishnan, A., & Figliozzi, M. (2019). Maximum coverage capacitated facility location problem with range constrained drones. *Transportation Research Part C: Emerging Technologies, 99*, 1–18.

Claesson, A., Fredman, D., Svensson, L., Ringh, M., Hollenberg, J., Nordberg, P., & Ban, Y. (2016). Unmanned aerial vehicles (drones) in out-of-hospital-cardiac-arrest. *Scandinavian Journal of Trauma, Resuscitation and Emergency Medicine, 24*(1), 124.

Claesson, A., Svensson, L., Nordberg, P., Ringh, M., Rosenqvist, M., Djarv, T., & Hollenberg, J. (2017). Drones may be used to save lives in out of hospital cardiac arrest due to drowning. *Resuscitation, 114*, 152–156.

de Oliveira Silva, L., de Mello Bandeira, R. A., & Campos, V. B. G. (2017). The use of UAV and geographic information systems for facility location in a post-disaster scenario. *Transportation Research Procedia, 27*, 1137–1145.

Dormehl, L. (2018). Intel wants its fleet of drones to monitor America's aging, unsafe bridges. *Digital Trends*. Retrieved February 24, 2019, from https://tinyurl.com/tmcx9qk

Durrant-Whyte, H., & Bailey, T. (2006). Simultaneous localization and mapping: Part I. *IEEE Robotics & Automation Magazine, 13*(2), 99–110.

Dwyer, F. (2019). Drones for good: UAVs help survey poorly mapped areas of Africa. *WeTalkUAV*. Retrieved August 12, 2019, from https://tinyurl.com/ycj9lhb8

Fahlstrom, P. G., & Gleason, T. J. (2012). *Introduction to UAV systems* (4th ed.). Chichester, UK: Wiley.

Faust, A., Palunko, I., Cruz, P., Fierro, R., & Tapia, L. (2017). Automated aerial suspended cargo delivery through reinforcement learning. *Artificial Intelligence, 247*, 381–398.

Feng, Q., Liu, J., & Gong, J. (2015). Urban flood mapping based on unmanned aerial vehicle remote sensing and random forest classifier—A case of Yuyao, China. *Water, 7*(4), 1437–1455.

Fuller, S. L. (2017). A day in the life of a US air force drone pilot. *Avionics International*. Retrieved from https://tinyurl.com/y74p53to

Gadi, V. K., Tang, Y. R., Das, A., Monga, C., Garg, A., Berretta, C., & Sahoo, L. (2017). Spatial and temporal variation of hydraulic conductivity and vegetation growth in green infrastructures using infiltrometer and visual technique. *Catena, 155*, 20–29.

Gaitani, N., Burud, I., Thiis, T., & Santamouris, M. (2017). High-resolution spectral mapping of urban thermal properties with Unmanned Aerial Vehicles. *Building and Environment, 121*, 215–224.

Gevaert, C. M., Persello, C., Sliuzas, R., & Vosselman, G. (2017). Informal settlement classification using point-cloud and image-based features from UAV data. *ISPRS Journal of Photogrammetry and Remote Sensing, 125*, 225–236.

Goodchild, A., & Toy, J. (2018). Delivery by drone: An evaluation of unmanned aerial vehicle technology in reducing CO2 emissions in the delivery service industry. *Transportation Research Part D: Transport and Environment, 61*, 58–67.

Gu, X., Abdel-Aty, M., Xiang, Q., Cai, Q., & Yuan, J. (2019). Utilizing UAV video data for in-depth analysis of drivers' crash risk at interchange merging areas. *Accident Analysis & Prevention, 123*, 159–169

Haidari, L. A., Brown, S. T., Ferguson, M., Bancroft, E., Spiker, M., Wilcox, A., & Lee, B. Y. (2016). The economic and operational value of using drones to transport vaccines. *Vaccine, 34*(34), 4062–4067.

Hassaan, O., Nasir, A. K., Roth, H., & Khan, M. F. (2016). Precision forestry: Trees counting in urban areas using visible imagery based on an unmanned aerial vehicle. *IFAC-PapersOnLine, 49*(16), 16–21.

Hemmelder, S., Marra, W., Markies, H., & De Jong, S. M. (2018). Monitoring river morphology bank erosion using UAV imagery–A case study of the river Buëch, Hautes-Alpes, France. *International Journal of Applied Earth Observation and Geoinformation, 73*, 428–437.

Hong, I., Kuby, M., & Murray, A. T. (2018). A range-restricted recharging station coverage model for drone delivery service planning. *Transportation Research Part C: Emerging Technologies, 90*, 198–212.

Inzerillo, L., Di Mino, G., & Roberts, R. (2018). Image-based 3D reconstruction using traditional and UAV datasets for analysis of road pavement distress. *Automation in Construction, 96*, 457–469.

Kanistras, K., Martins, G., Rutherford, M. J., & Valavanis, K. P. (2015). Survey of unmanned aerial vehicles (UAVs) for traffic monitoring. In K. Valavanis & G. Vachtsevanos (Eds.), *Handbook of unmanned aerial vehicles*. Dordrecht: Springer.

Karpowicz, J. (2018). Stereographic mapping and drones: Taking UAS mapping to the limit. *Commercial UAV News*. Retrieved November 14, 2018, from https://tinyurl.com/y7bbt9wu

Katz, D. (2018). Surveyors and other professional drone operators should be focused on accuracy and reliability. *Commercial UAV News*. Retrieved March 17, 2019, from https://tinyurl.com/y7vkknn5

Khan, M. A., Ectors, W., Bellemans, T., et al. (2017). Unmanned aerial vehicle–based traffic analysis: Methodological framework for automated multivehicle trajectory extraction. *Transportation Research Record, 2626*, 25–33.

Khan, M. A., Ectors, W., Bellemans, T., et al. (2018). Unmanned aerial vehicle-based traffic analysis: A case study for shockwave identification and flow parameters estimation at signalized intersections. *Remote Sensing, 10*(3), 458.

Larsen, M. P., Eisenberg, M. S., Cummins, R. O., & Hallstrom, A. P. (1993). Predicting survival from out-of-hospital cardiac arrest: A graphic model. *Annals of Emergency Medicine, 22*(11), 1652–1658.

Liang, H., Li, W., Zhang, Q., Zhu, W., Chen, D., Liu, J., & Shu, T. (2017). Using unmanned aerial vehicle data to assess the three-dimension green quantity of urban green space: A case study in Shanghai, China. *Landscape and Urban Planning, 164*, 81–90.

Lillian, B. (2018). Urban BVLOS drone project completed at cemetery site. *Unmanned Aerial Online*. Retrieved January 23, 2019, from https://tinyurl.com/yab7ldfk

Lin, Y., Jiang, M., Yao, Y., Zhang, L., & Lin, J. (2015). Use of UAV oblique imaging for the detection of individual trees in residential environments. *Urban Forestry & Urban Greening, 14*(2), 404–412.

Manfreda, S., McCabe, M., Miller, P., Lucas, R., Pajuelo Madrigal, V., Mallinis, G., & Müllerová, J. (2018). On the use of unmanned aerial systems for environmental monitoring. *Remote Sensing, 10*(4), 641.

Margaritoff, M. (2018). Kespry's new thermal drone inspection can accurately assess specific building damage. *The Drive*. Retrieved from https://tinyurl.com/y9xmt4or

Matsunaga, K. (2017). Prodrone develops a long-distance delivery flight helicopter "DELIX". *Prodrone*. Retrieved from https://tinyurl.com/y7epqk9s

Murray, C. C., & Chu, A. G. (2015). The flying sidekick traveling salesman problem: Optimization of drone-assisted parcel delivery. *Transportation Research Part C: Emerging Technologies, 54*, 86–109.

Nex, F., & Remondino, F. (2014). UAV for 3D mapping applications: a review. *Applied geomatics, 6*(1), 1–15.

Näsi, R., Honkavaara, E., Blomqvist, M., Lyytikäinen-Saarenmaa, P., Hakala, T., Viljanen, N., & Holopainen, M. (2018). Remote sensing of bark beetle damage in urban forests at individual tree level using a novel hyperspectral camera from UAV and aircraft. *Urban Forestry & Urban Greening, 30*, 72–83.

Natural England. (2010). Natural England's Green Infrastructure Guidance (NE176). Retrieved June 12, 2019, from http://publications.naturalengland.org.uk/publication/35033

Park, S. I., & Um, J. S. (2018). Differentiating carbon sinks versus sources on a university campus using synergistic UAV NIR and visible signatures. *Environmental Monitoring and Assessment, 190*(11), 652.

Pulver, A., & Wei, R. (2018). Optimizing the spatial location of medical drones. *Applied Geography, 90*, 9–16.

Pulver, A., Wei, R., & Mann, C. (2016). Locating AED enabled medical drones to enhance cardiac arrest response times. *Prehospital Emergency Care, 20*(3), 378–389.

Rabta, B., Wankmüller, C., & Reiner, G. (2018). A drone fleet model for last-mile distribution in disaster relief operations. *International Journal of Disaster Risk Reduction, 28*, 107–112.

Rakha, T., & Gorodetsky, A. (2018). Review of Unmanned Aerial System (UAS) applications in the built environment: Towards automated building inspection procedures using drones. *Automation in Construction, 93*, 252–264.

Rattigan, K. (2019). Shell launches drone inspection trial. *National Law Review*. Retrieved from https://tinyurl.com/ybghajss

Rees, M. (2018). Leonardo target drone authorised for military operations. *Unmanned Systems Technology*. Retrieved April 13, 2019, from https://tinyurl.com/yd23jn9s

Rosenfeld, A. (2019). Are drivers ready for traffic enforcement drones? *Accident Analysis & Prevention, 122*, 199–206.

Rossi, M., Brunelli, D., Adami, A., Lorenzelli, L., Menna, F., Remondino, F. (2014). Gas-drone: Portable gas sensing system on UAVs for gas leakage localization. In *Sensors* (p. 1434). IEEE.
Rutkin, A. (2016). Blood delivered by drone. *NewScientist, 232*, 24.
Salvo, G., Caruso, L., & Scordo, A. (2014). Urban traffic analysis through an UAV. *Procedia: Social and Behavioral Sciences, 111*, 1083–1091.
Sawyer, T. (2018). 60-mile-long drone inspection flight points to the future. *Engineering News-Record*. Retrieved December 9, 2018, from https://tinyurl.com/yccfxo7r
Seguin, C., Blaquière, G., Loundou, A., Michelet, P., & Markarian, T. (2018). Unmanned aerial vehicles (drones) to prevent drowning. *Resuscitation, 127*, 63–67.
Sinha, A., Tsourdos, A., & White, B. (2009). Monitoring the dispersion of a contaminant cloud in an urban region by a swarm of UAV sensors. *IFAC Proceedings Volumes, 42*(22), 7–12.
Tatum, M. C., & Liu, J. (2017). Unmanned aircraft system applications in construction. *Procedia Engineering, 196*, 167–175.
Terwilliger, B., Ison, D. C., Robbins, J., & Vincenzi, D. (2017). *Small unmanned aircraft systems guide: Exploring designs, operations, regulations, and economics*. Washington: Aviation Supplies & Academics.
Tomas, J. P. (2017). The role of drones in telecommunications tower inspection. *RCRWireless News*. Retrieved January 23, 2018, from https://tinyurl.com/y9fm3jfe
Tuna, G., Nefzi, B., & Conte, G. (2014). Unmanned aerial vehicle-aided communications system for disaster recovery. *Journal of Network and Computer Applications, 41*, 27–36.
Van Tilburg, C. (2017). First report of using portable unmanned aircraft systems (drones) for search and rescue. *Wilderness & Environmental Medicine, 28*(2), 116–118.

Chapter 3
A Primer on Aeronautical Charts and Maps

Abstract The purpose of this chapter is to provide readers with a primer on aeronautical charts and maps. This is especially important for sUAS pilots and spotters as they evaluate the suitability of sites for field operations. Although space limitations will prevent us from detailing all the nuances of the visual flight rules (VFR) charts from the Federal Aviation Administration (FAA), we provide readers with enough background to better understand the geographic constraints with sUAS operations. This includes information on airspace classifications and sUAS restrictions, control areas, special airspace and/or restricted airspace, local terrestrial obstacles, in-air hazards, military training routes, terminal charts, and terrain.

Keywords Visual flight rules · Aeronautical chart · Airspace · Sectional charts · Instrument flight rules · Above ground level · Mean sea level

3.1 Introduction

One of the core challenges associated with the operation of small unmanned aerial systems is safety. As will be detailed in Chap. 4, there are scores of safety considerations to be considered both prior to and during sUAS operation. Many of these safety guidelines are formalized by the Federal Aviation Administration (FAA) under the Part 107 guidelines (FAA 2016). Part 107 includes basic rules for maintaining visual line of sight (VLOS) with the sUAS at all times, not flying the sUAS over any persons not directly participating in the operations, requirements for daylight-only operations, making sure the sUAS stays below 400 ft above ground level (AGL), obtaining air traffic control permission to operate in Class B, C, D, and E airspace, and many other stipulations that remote pilots need to be aware of. As mentioned in Chap. 2, although some of these requirements can be waived, there is a process associated with such requests. Again, these details are covered more thoroughly in Chap. 4.

One of the most indispensable tools for improving operational safety of sUAS operations is the use of the FAA visual flight rules (VFR) chart(s) (Fig. 3.1). VFR charts were originally designed to assist pilots flying an aircraft without using instrumentation (e.g., distance measuring instruments, radio beacons, GPS).

T. H. Grubesic, J. R. Nelson, *UAVs and Urban Spatial Analysis*,
https://doi.org/10.1007/978-3-030-35865-5_3

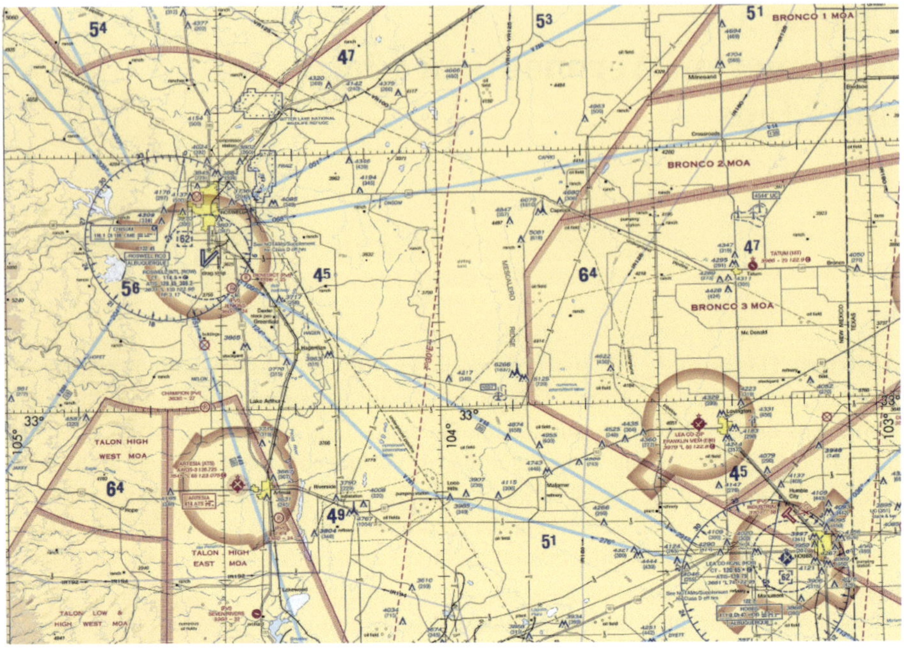

Fig. 3.1 FAA visual flight rules (VFR) chart

In short, VFR flights require pilots to see outside the cockpit and control the aircraft's height, direction, speed, and safely. VFR pilots must be able to "see and avoid" obstacles in the air (i.e., other aircraft), but also terrestrial obstacles such as power lines, geologic formations, buildings, and the like. Early versions of the VFR charts were cobbled together by individual pilots using personal notes to help them navigate between origins and destinations. This included sketches of landing sites, local obstacles, or other significant features on a given route (ICAO 2019). In time, both pilots and ground control agencies realized that standardized products and symbols were needed to support both domestic and international air travel (ICAO 2019). Between 1919 and 1949, significant advancements in chart symbology, style, required elements, and broad coverage were achieved. Today, the VFR charts are completely standardized and updated at regular intervals by the FAA.

The purpose of this chapter is to provide readers with a primer on aeronautical charts and maps, paying particular attention to the FAA VFR charts. This is especially important for sUAS pilots and spotters as they evaluate the suitability of sites for field operations. Although space limitations will prevent us from detailing all the nuances of the VFR charts, we provide readers with enough background to better understand the geographic constraints with sUAS operations. This includes information on airspace classifications and sUAS restrictions, control areas, special airspace and/or restricted airspace, local terrestrial obstacles, in-air hazards, military training routes, terminal charts, and terrain. We also want to emphasize to readers that the information provided here is for reference purposes, only. Please check your local rules and regulations prior to launching an sUAS anywhere. As the FAA would say, "know before you fly."

3.2 Airspace

Prior to a more in-depth exploration of the VFR charts, it is important that readers of this book understand the basic composition of the US national airspace system. Figure 3.2 highlights the structure of the system. Each of the classes, A, B, C, D, E, and G are structured to facilitate the safe movement through these spaces, ensuring usability for a wide range of aircraft.

3.2.1 Class A Airspace

Class A Airspace consists of all areas between 18,000 ft mean sea level (MSL) and 60,000 ft MSL in the United States. Class A also extends 12 nautical miles off the US coastline. All operations in this space must be conducted under instrument flight rules (IFR), unless otherwise authorized. To be clear, this is not an area that anybody should be operating an sUAS.

3.2.2 Class B Airspace

Class B airspace belongs to some of the busiest airports in the United States—including Phoenix (PHX), Chicago (ORD), San Francisco (SFO), Dallas-Fort Worth (DFW), and many others. The skies are so busy around these airports that any aircraft flying through Class B airspace is required to have an active transponder so

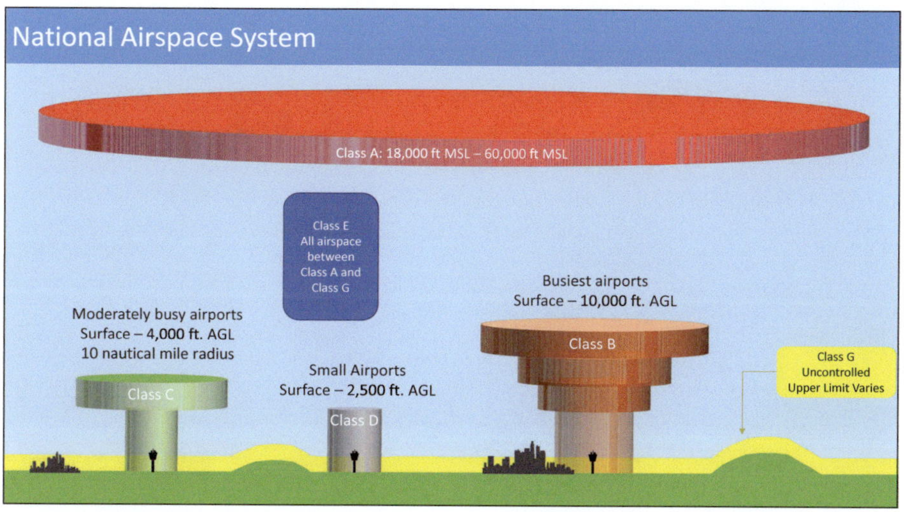

Fig. 3.2 National airspace system

that air traffic control (ATC) can instantly identify its altitude when it gets within 30 nautical miles of the airport. This is called Mode C Veil. All airports operating with Class B airspace have these altitude reporting transponders. As detailed by the FAA (2008), the shape of Class B airspaces is bespoke to each airport. It consists of multiple areas (i.e., layers) that extend to an altitude of 10,000 MSL (Fig. 3.2). The structure of these layers is often referred to as an "upside down wedding cake," but we think it looks more like a tiered funnel. Either way, sUAS pilots are primarily concerned with the bottom tier (closest to the ground) because this is the airspace that could (potentially) be used for field operations. Although there are variations by airport, Class B airspace generally extends 7 nautical miles from the center of an airport, in all directions (14 nautical miles in diameter). sUAS pilots must get ATC clearance and FAA authorization to work in this airspace. Obviously, understanding the geographic footprint of this airspace is really important, and we will discuss how this is done later in the chapter when exploring VFR charts. Figure 3.3 highlights the geographic footprint of Class B airspace for the Southwestern United States. Notice that no two airports have an identical shape for their Class B airspace.

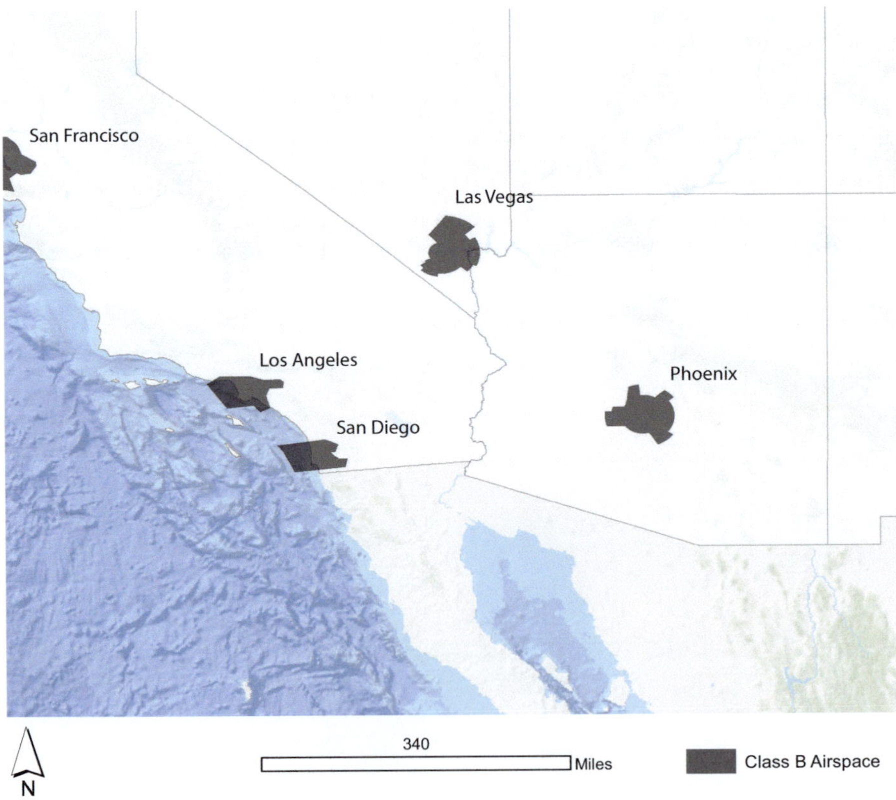

Fig. 3.3 Geographic footprint of class B airspace in the Southwestern United States

3.2.3 Class C Airspace

Class C airspace generally consists of areas from the surface to 4000 ft above the airport elevation (charted in MSL) (Fig. 3.2). Much like Class B airspace, Class C resembles a tiered funnel. The layer closest to the surface extends 5 nautical miles from the center of the airport in all directions (10 statute miles in diameter). The top layer extends 10 nautical miles from the airport and includes altitudes between 1200 and 4000 ft. Class C airspace is not as busy as Class B, but do not let this fool you. Traffic in this airspace can be dense. Class C airspace includes many relatively large airports, such as Albuquerque (ABQ), Austin (AUS), San Antonio (SAT), Columbus (CMH), and Indianapolis (IND), among others. Figure 3.4 displays areas with Class C airspace for the Southwestern United States.

Notice how things start to get complicated in parts of Southern California, where Class C airspace is immediately proximal to Class B around Los Angeles (LAX). For sUAS operations, Part 107 dictates that all operations within Class C airspace require FAA authorization, including a waiver or permission from ATC. Please note,

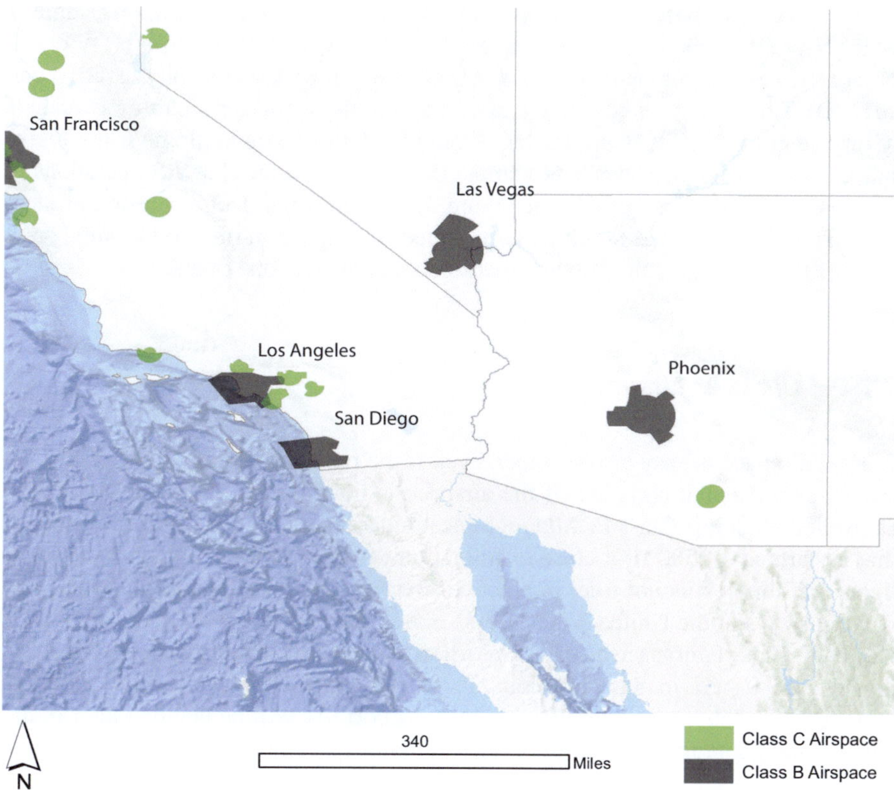

Fig. 3.4 A visual comparison of class B and C airspace in the Southwestern United States

when flying in these busier locales, it is always wise to notify ATC prior to any operations. They are often aware of additional information regarding the airspace at any given time (e.g., schedule military operations) that may impact sUAS flights.

3.2.4 *Class D Airspace*

Class D airspace does not look like a wedding cake or layered funnel, so the party is over. It consists of airspace from the surface to 2500 ft above the airport elevation (charted in MSL) (Fig. 3.2). Class D airspace is always connected to airports that have an operational control tower, but that tower may not always be staffed or operating. In these instances, Terwilliger et al. (2017) note that the airport may revert to either Class E or Class G operations. If the ATC is operating and the Class D airspace is active, sUAS operations require FAA permission or a waiver. Many of the airports operating with Class D airspace are considered "reliever airports." As detailed by Grubesic et al. (2017), reliever airports must provide general aviation access to the surrounding area, have 100 or more based aircraft, or have 25,000 annual itinerant operations. In the Phoenix, AZ metropolitan area, there is an entire network of reliever airports, some of which have many operations. For example, Deer Valley Municipal Airport (DVT) had 421,161 tower operations in 2018 (ATADS 2018). In fact, as recently as 2010 and 2012, DVT ranked as one of the top 30 airports in the world for aircraft movements (Grubesic et al. 2017). Again, readers should keep this in mind. Some Class D airspace is extraordinarily busy, and without the proper clearance, sUAS operations in these areas could create problems. Figure 3.5 displays the geographic footprint of Class D airspace for the Southwestern United States. Again, notice the number of Class D locales ringing the Phoenix metropolitan area—a total of six.

3.2.5 *Class E Airspace*

Class E airspace is easy to remember, even if the operational conditions of the airspace are somewhat complex. If the airspace is not Class A, B, C, or D, but it is controlled, then it is Class E. Although most federal sources (i.e., the FAA) suggest that it starts at 14,500 ft; a close examination of most VFR sectional charts highlights that this just is not true, in practice. Because there are so many airports and victor airways in the United States, Class E airspace usually starts at 1200 ft AGL.[1] However, Class E airspace also has transition areas, which means the airspace drops down to 700 ft AGL in many locations. This step down is designed to protect aircraft on their approach or departure from a local airport. As will be detailed later in the chapter, the geographic footprint of Class E airspace can be quite complex, almost

[1]Victor airways are flight paths designated by the FAA to accommodate aircraft VFR activity from 1200 ft AGL to altitudes up to but not including 18,000 ft above MSL (FAA 2008).

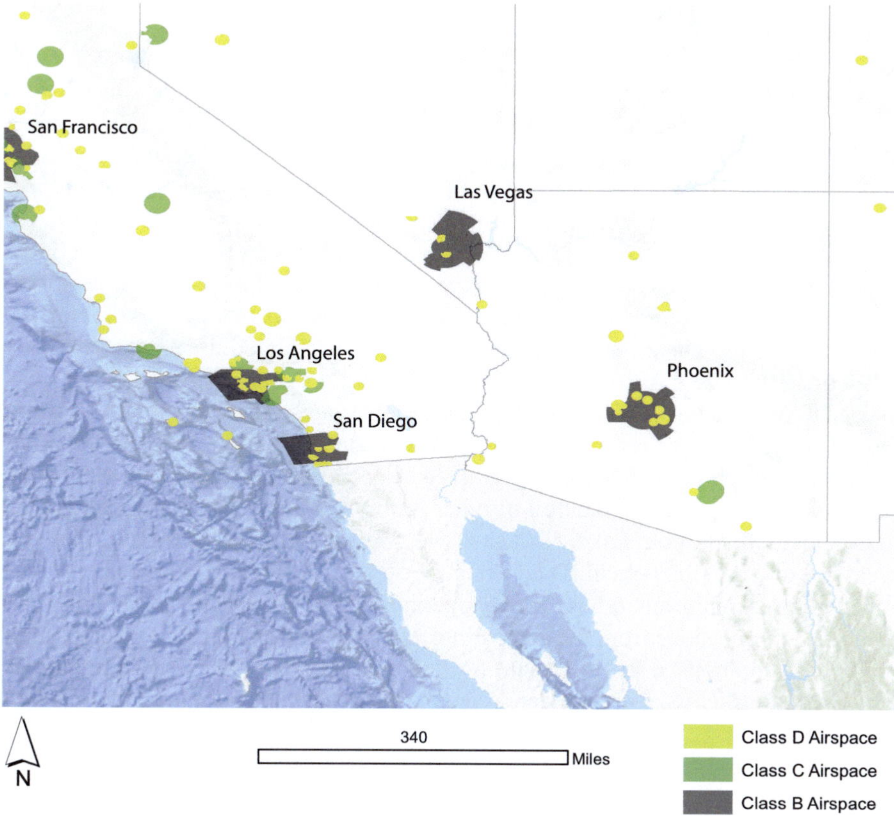

Fig. 3.5 A visual comparison of class B, C, and D airspace in the Southwestern United States

looking amoeboid for some areas. There may be multiple airports in the area operating with Class E airspace, but again, it is structured to protect aircraft during approach or departure. To make things more confusing, some class E airspace is structured like a layered funnel (akin to Class B or C airspace), extending all the way down to the surface—but it is not as tightly controlled as Class B or C. For sUAS operations, ATC permission or a FAA airspace waiver is required when Class E airspace extends to the surface.

3.2.6 Class G Airspace

Class G airspace is a catch-all. Basically, it is all airspace not designated as Class A, B, C, D, or E. This is important airspace for sUAS operators because it is uncontrolled. It usually extends from the surface to 700 or 1200 ft AGL. As noted by the FAA (2018), although ATC has no authority or mandate to control air traffic in these areas, visual flight rule minimums still apply.

3.2.7 Special Use Airspace

Special use airspace is scattered throughout the United States and generally consists of the following categories: (1) prohibited areas, (2) restricted areas, (3) warning areas, (4) military operation areas, (5) alert areas, and (6) controlled firing areas.

Perhaps the first and only thing that sUAS pilots need to understand about *prohibited areas* is that they are completely off-limits for drone operations. Later in the chapter, we will highlight exactly how to identify these areas on the VFR sectional charts. For now, it is interesting to note that the United States has a handful of permanent prohibited areas. All make sense, but it is an odd mix of locations (Table 3.1). For example, the Disney theme parks lobbied for prohibited area status after the attacks of 9/11. Then, in 2014, Disney wanted to add sUAS that carried LED screens and small pyrotechnics for their nightly light shows and seasonal fireworks displays. However, because the park had prohibited area status, this was prohibited. After 2 years of lobbying, the FAA finally granted a special exception for the no-fly zone (Lumb 2016). There are a handful of other prohibited areas throughout the United States, including Groom Lake, NV (Area 51), Merritt Island, FL (NASA), but these areas are so large, the typical ID tags (e.g., P-50) for VFR charts do not apply.

There are conditions where sUAS operations can be conducted in *restricted areas*, but permissions from the controlling agencies are rare. Restricted areas generally contain airspace that represent a significant danger to non-participating aircraft, including sUAS. Restricted areas denoted on VFR charts with an "R" followed by a number (e.g., R-2307) are typically associated with military bases and the defense industrial complex. For example, the Laguna Army Airfield near Yuma, AZ (R-2307) is a very large restricted area, home to the Marine Aviation Weapons and Tactics Squadron 1 (MAWTS-1). The restricted area itself offers a massive amount of open terrain for the testing of air-to-ground weapons. This is not somewhere you should be flying an sUAS.

Although *warning areas* have many things in common with restricted areas, they are not exactly equivalent. Warning areas extend from 12 nautical miles outward from the US coastline and contain activity that may be hazardous to non-participating aircraft. Similar to restricted areas, these activities include the use of artillery and

Table 3.1 Special use airspace: prohibited areas

Location	ID	Notes
Thurmont, MD	P-40	Camp David (Presidential Retreat)
Kennebunkport, ME	P-67	Bush Family Compound
Washington, DC	P-56	U.S. Capitol, White House, Naval Observatory
Mount Vernon, VA	P-73	Home of George Washington
Kingsland, GA	P-50	Naval Submarine Base Kings Bay
Crawford, TX	P-49	Bush Family Ranch
Amarillo, TX	P-47	Pantex Nuclear Assembly Plant
Northern Minnesota	P-204, 205, 206	Boundary Waters Canoe Area Wilderness
Silverdale, WA	P-51	Naval Base Kitsap
Various		Disney Theme Parks

ammunition, military aircraft operating at high speeds, and the use of electronic attack equipment. Warning areas are denoted on VFR charts with a "W" followed by a number (e.g., W-93). In this instance, W-93 North corresponds to the "Control Area Pacific Low," off the coast of Gold Beach, OR. Again, where sUAS operators are concerned, unless you are launching and operating a drone off the deck of a ship, these are areas that need to be avoided and are well-beyond VLOS from the shoreline.

Airspace with defined vertical and lateral limits for isolating military training activity from IFR traffic are known as *military operation areas* (MOAs). Interestingly, sUAS activity is permitted in MOAs. However, sUAS operators always need to be aware of military air traffic in these areas. MOAs are not numbered, but sectional charts will include information on the times of operation, altitudes affected, and the controlling agency. Prior to sUAS operations, it would be wise to determine exactly what might be going on in the MOA, that day. The Pinon Canyon MOA, located outside of Trinidad, Colorado is a good exemplar. It is a massive area (955 km^2) that serves as the training site for Fort Carson. This includes the fourth Combat Aviation Brigade, which operates the Gray Eagle drone for both reconnaissance and security (Roeder 2018). Until recently, all training and use of the Gray Eagle drones occurred within the bounds of Fort Carson. However, the post is asking the FAA to create a corridor between Fort Carson and the Pinon Canyon Maneuvers site to provide pilots with more experience with longer-distance flights (Roeder 2018).

Activity within *alert areas* generally corresponds to high volumes of pilot training or unusual aerial activity. They are denoted on VFR charts with an "A" followed by a number (e.g., A-231). In this instance, A-231 corresponds to an area northwest of Phoenix, AZ associated with Luke Air Force Base that hosts concentrated student jet transition training. For sUAS operators, it is important to remember that while drone operations are permitted in these areas, all activities must be conducted in accordance with the existing regulations. Again, this is one of those situations where it would be important to contact a local flight service station to inquire about activity in the area prior to any sUAS operation.

One last set of areas to be concerned with an sUAS pilot are *controlled firing areas* (CFAs), which are locations where people blow up stuff. The problem with identifying CFAs is that they are not charted and there is no controlling agency in charge of them. For a typical aircraft, these types of activities are largely irrelevant because they occur at ground level and do not pose a threat to aircraft flying overhead. However, for sUAS operators, there is a chance (albeit small) that the dust and debris generated by controlled firing activity could impact sUAS sensors, potentially leading to a fly-away or, worse, a crash. Pro-tip, try to avoid areas where people are doing this kind of thing. Yes, the sUAS footage of things exploding would be cool to watch, but know that operations in these areas could be problematic for the drone.

Finally, it is incumbent upon every sUAS operator to understand that there are a variety of additional no-fly areas. As a quick refresher, do not fly anywhere near an airport, a heliport, major league sports stadium or racetrack, a wildfire, a nuclear power plant, in any of the prohibited or restricted airspaces detailed above, within a national park or near an NOAA marine protected area. We encourage readers to become familiar with the Know Before You Fly site and application from the FAA (http://knowbeforeyoufly.org/). Also, if you have not been exposed to the "Buzzy the Drone" video series, you might consider checking them out. Of course, we are assuming readers of this book like to draw cartoon faces and craft bad poetry to entertain their friends. No? Then we apologize for the suggestion to visit Buzzy.

3.3 Sectional Charts

Some readers might find it odd that we are dedicating a large subsection of this book to reading and interpreting VFR sectional charts. This is likely the case with most geographers and any other readers with a disciplinary background that involves mapping and spatial analysis. Yes, we know that you can read maps. However, word to the wise, you might know less than you think, especially when it comes to reading VFR charts. Not only are they extraordinarily complex, they are custom built for a specific application—flying with few instruments. So, unless you do this regularly, or have had coursework on air navigation, Sect. 3.3 might prove helpful. Further, many readers of this book likely have (or will soon) their drone pilot license, which is great, but given the sheer number of questions on the Part 107 test that use VFR charts, a refresher on the reading and interpretation of these charts is probably helpful, especially if the time to renew your license is creeping closer. Finally, VFR sectional charts are the single best piece of information available to sUAS pilots when determining the requirements and restrictions for work in the field. Think of this as a free insurance policy—if the ability to read VFR charts saves you the time and expense associated with driving to a field site where you cannot fly, great! More importantly, if it saves you from operating in a restricted area where you could be fined (or worse), even better. We cannot hope to cover all the nuances of the VFR charts, but we cover the most important elements for sUAS pilots and spotters. All the VFR charts are available for download from the FAA (https://tinyurl.com/ybah5yun). Reader beware, throughout this section, we annotate maps with bright red arrows, lines, and boxes—this is not a color normally used in the VFR charts.

3.3.1 Airspace on VFR Charts

As detailed in Sect. 3.2, the composition of the US national airspace system is extremely diverse. Understanding the vertical and horizontal nature of airspace is critical for sUAS mission planning and execution. Because sUAS pilots do not need to be concerned with Class A (Alpha) airspace (areas between 18,000 and 60,000 ft MSL), we shift focus to Class B (Bravo) airspace. Figure 3.6 displays the VFR chart for the region in/around Denver, CO. There is a lot going on here, so let us start with the basics. First, Class B airspace is denoted with the solid blue lines. In Fig. 3.6, the [*Item1*] label is sitting directly on top of the Bravo line. Notice the plurality of these lines. This simply corresponds to the various shelves associated with the airspace

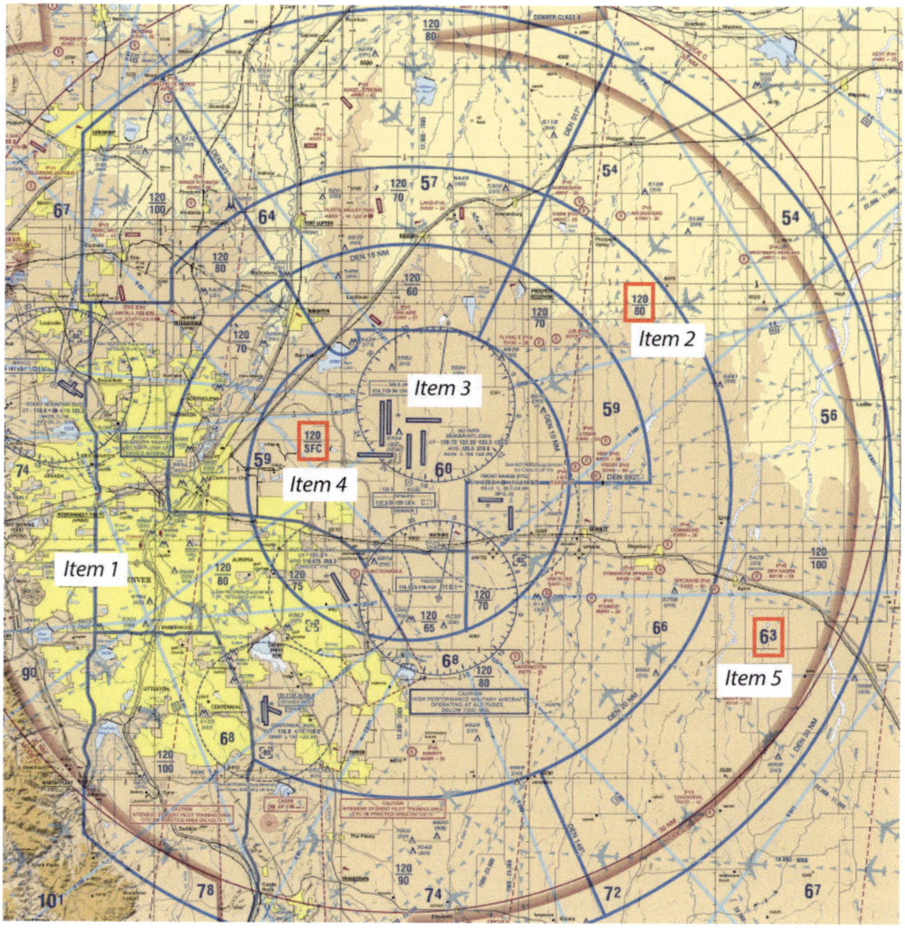

Fig. 3.6 Class B airspace in the Denver metropolitan area

(remember the wedding cake/tiered funnel example). Further, although the lines appear to adhere to a concentric shape on the eastern side of Denver, this is not the case as one moves toward the west, where the airspace abuts the Rocky Mountains. [*Item2*] provides guidance on the altitude associated with the Bravo shelf for that area. In this instance, Bravo airspace extends from 8000 to 12,000 ft (markers represent hundreds of feet). As one moves closer to the DEN airport [*Item3*], the height of the Bravo shelves changes. For example, [*Item4*] denotes that Class B airspace extends from the surface (SFC) to 12,000 ft. Finally, [*Item5*] corresponds to a maximum elevation figure (MEF) for a quadrangle. This is a measure of the highest feature within the area. Although this is not particularly relevant for sUAS operators, it is critical for aircraft pilots, helping them to avoid mountain peaks or tall towers in their flight path.

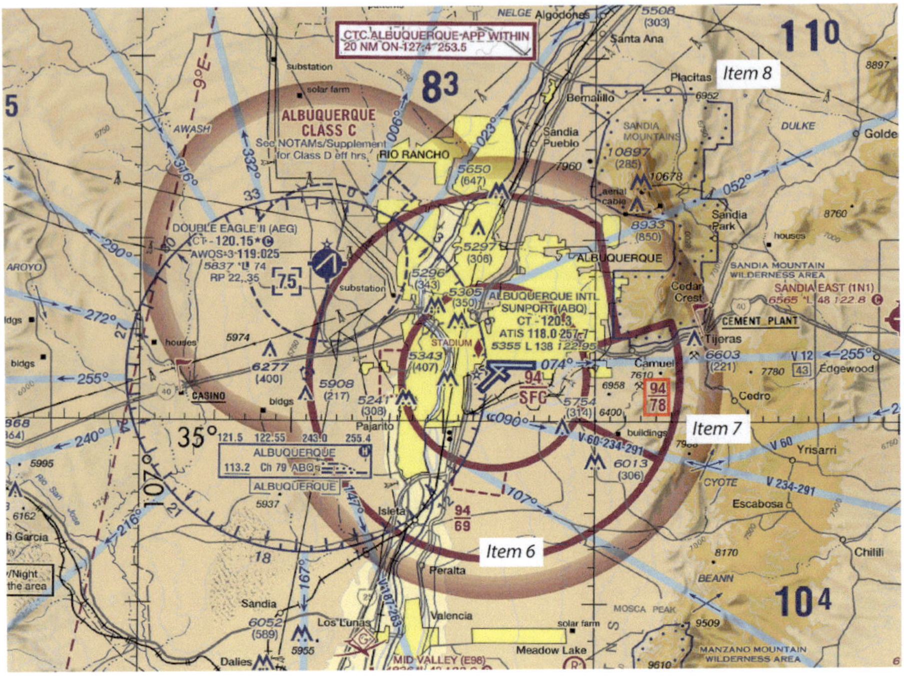

Fig. 3.7 Class C airspace in the Albuquerque metropolitan area

Figure 3.7 displays the VFR chart for the area in/around the Albuquerque International Sunport (ABQ), which is Class C (Charlie) airspace and is represented by a solid magenta line on the VFR sectional. The [*Item6*] marker is sitting directly on top of the Charlie line. Once again, there is more than one of these lines, denoting the multiple shelves associated with Class C airspace around ABQ. On the eastern side of the Charlie airspace, [*Item7*] provides guidance on the altitude, which extends from 7800 to 9400 ft. As one moves closer to the airport, Class C extends from the surface to 9400 ft. Notice how high some of the MEF markers are for the area immediately east of Albuquerque (e.g., 11,000 ft, 10,400 ft). This is the Sandia Mountain Range. Finally, [*Item8*] marks the village of Placitas, NM, which sits outside of any/all restrictions connected to the ABQ airport. It is important to acknowledge that ABQ is surrounded by Class E airspace (the thicker, diffuse magenta line) that starts at 700 ft (within the line) and 1200 ft outside of the line. We will continue the discussion on Class E airspace below.

Figure 3.8 displays Class D (Delta) airspace in/around the Klamath Falls, OR airport (LMT), which is represented by the dashed blue line on the VFR sectional. The marker for [*Item9*] sits right on top of the circular Delta line. Readers should take note of the second blue line with hatch marks [*Item10*]. This does *not* demarcate the Class D airspace. Rather, it highlights that the airport operates a very high frequency (VHF) omni-directional range (VOR) radio navigation system for aircraft. Basically, it allows aircraft to determine its position, relative to a network of fixed radio beacons around the airport. It is also important to highlight [*Item11*],

Fig. 3.8 Class D and Class E airspace in the Klamath falls area

which directs the readers to "See NOTAMS/Supplement for Class D eff hrs." There are two important elements to decompose here. NOTAM is an acronym for "notice to airmen" and is information relevant to operations for an airport (FAA 2008). In this case, the NOTAM informs readers of the VFR chart that the control tower for LMT operates between the hours of 0700 and 2200 (7 am to 10 pm). For sUAS pilots, this means that prior to conducting an operation within Delta airspace, one could (and should) reach out to ATC for clearance, keeping in mind that the tower is not always staffed. Finally, [*Item12*] represents the upper limit of the Class D airspace, at 6600 ft. Remember, Delta airspace usually extends from the surface to 2500 MSL. Because the Klamath Falls airport is situated at 4094 ft MSL, adding 2500 ft for the Delta airspace (4094 + 2500 = 6594) yields a value of ~6600 MSL.

Class E (Echo) surface areas are also displayed in Fig. 3.8, represented by thicker, diffuse magenta line [*Item13*]. Recall that Echo starts at 700 or 1200 ft AGL. Inside the magenta line, it starts at 700 ft, and outside of the magenta line, it starts at

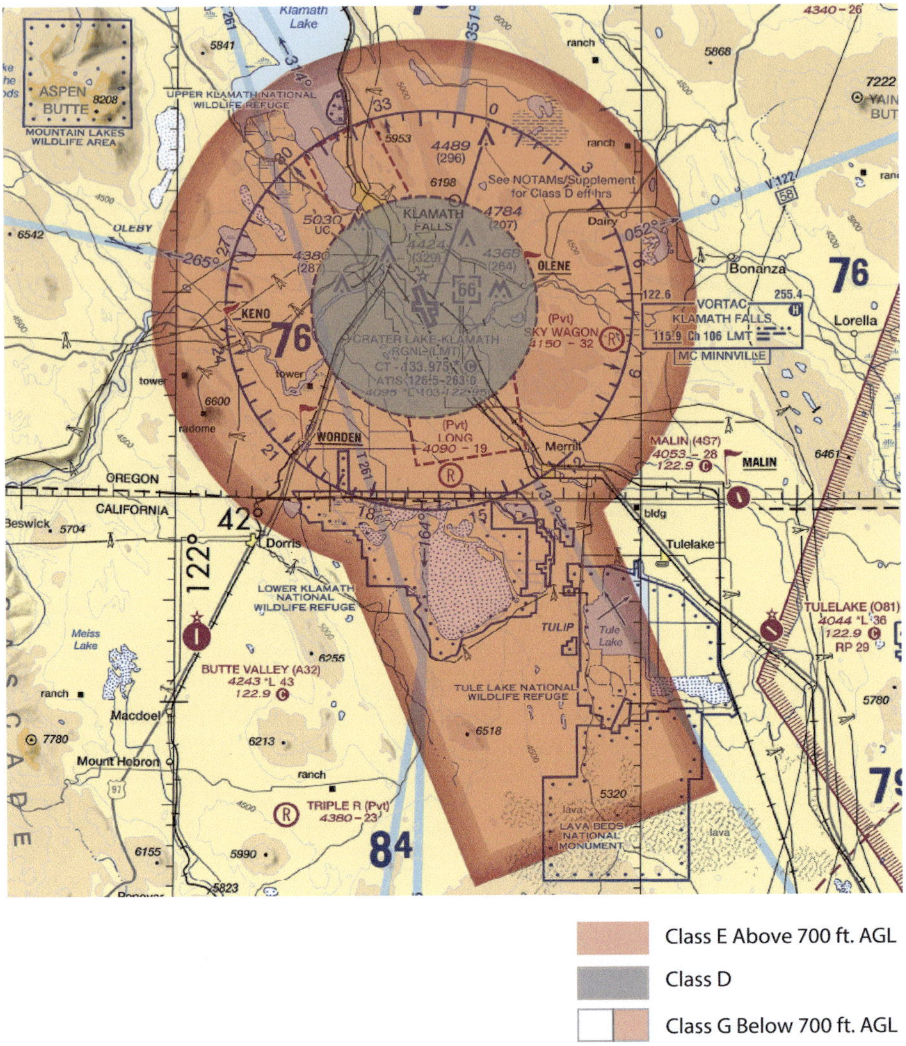

Fig. 3.9 Class D and Class E airspace in the Klamath falls area

1200 ft. The Echo transition space in Fig. 3.8 is irregularly shaped, circling the LMT airport, but also extending to the southeast, to help protect runway approach and departure procedures. The airspace below 700 ft, within the diffuse magenta line, but outside of the Class D airspace, is considered **Class G**. This is also true for all the areas outside of the Echo transition airspace [*Item14*].

As a reader and sUAS operator, this information is critically important to digest. Class E and Class G airspace coexist geographically, but *not* altitudinally. Take a closer look at Fig. 3.9, which is a color-coded map of the airspace around LMT. The magenta area highlighted in Fig. 3.9 is considered Class E airspace from 700 ft

AGL, upward. This is *not* true for the gray area, which corresponds to the Class D airspace closer to the airport. Again, Delta starts at the surface and extends to 6600 ft MSL (2500 AGL). Echo airspace is everything else in maroon, above 700 ft. The confusing part is that Class G airspace *also* exists in the maroon area, extending from the surface to 699 ft. As an sUAS operator, you do not need any permissions to fly in Class G airspace, it is uncontrolled. So, if you are following Part 107 regulations and keeping the drone below 400 ft, no permissions are necessary in the magenta-colored areas or anywhere else outside of the defined Class E airspace (except for the MOA area denoted with red hatches) to the east. The remainder of the airspace in this portion of the sectional is uncontrolled, Class G.

3.3.2 Special Use Airspace on VFR Charts

Figure 3.10 is a composite graphic of the five different special use airspaces found on the VFR charts. The classic example of prohibited airspace is in Washington DC (Fig. 3.10a) and includes areas around the White House and the Naval Observatory [*Item15*]. Again, this is denoted with a blue, hatched pattern. Do not fly sUAS anywhere near these locations, unless you want to spend some time in federal prison. For readers interested in a captivating VFR chart, spend some time with the Las Vegas sectional. Not only does it include the Grand Canyon (*do not* fly your sUAS there), but it also includes Area 51. Figure 3.10b highlights the restricted area (R-4808 N) around Groom Lake, NV (Area 51), northwest of Las Vegas. Notoriously absent from the VFR chart is the extremely large runway complex located at Groom Lake (Fig. 3.10c) [*Item16*]. Although details are sparse, we do know that this is called Homey Airport (KXTA) (Marsh 2008), which plays host to Janet Airlines. Never heard of Janet? It is a small fleet of planes that serve employees commuting from Las Vegas International Airport to both Area 51 and the Tonopah Test Range. The morning commuter flight, which is an unmarked Boeing 737, leaves McCarran International (LAS) at 3:30 am. Although the official destination for these flights is always listed as the Tonopah Test Range, visual observers (and other people who have way too much time on their hands) have identified the actual destination as Groom Lake (XTA). Recent reports suggest that Janet operates between 13 and 16 roundtrip flights per day shuttling people to/from LAS and XTA (Warnica 2018). Figure 3.10d displays the Reveille Military Operation Area (red hatched pattern), which is linked to the Nellis Air Force Base complex in the Nevada desert. [*Item 17*], which is a stacked pair of parallel blue lines, denotes that Class E airspace starts at 9700 ft.

Figure 3.11a highlights the alert area (A-231) detailed earlier in this chapter. Located northwest of Phoenix, AZ, this is an area where student jet training is taking place, associated with Luke AFB. Readers should also take notice of the area denoted by white, with the thin blue hatched lines near the alert area. This is an area that requires ATC clearance prior to entry. Obviously, it also intersects with the alert area and the far eastern edge and also intersects with Bravo airspace. For sUAS

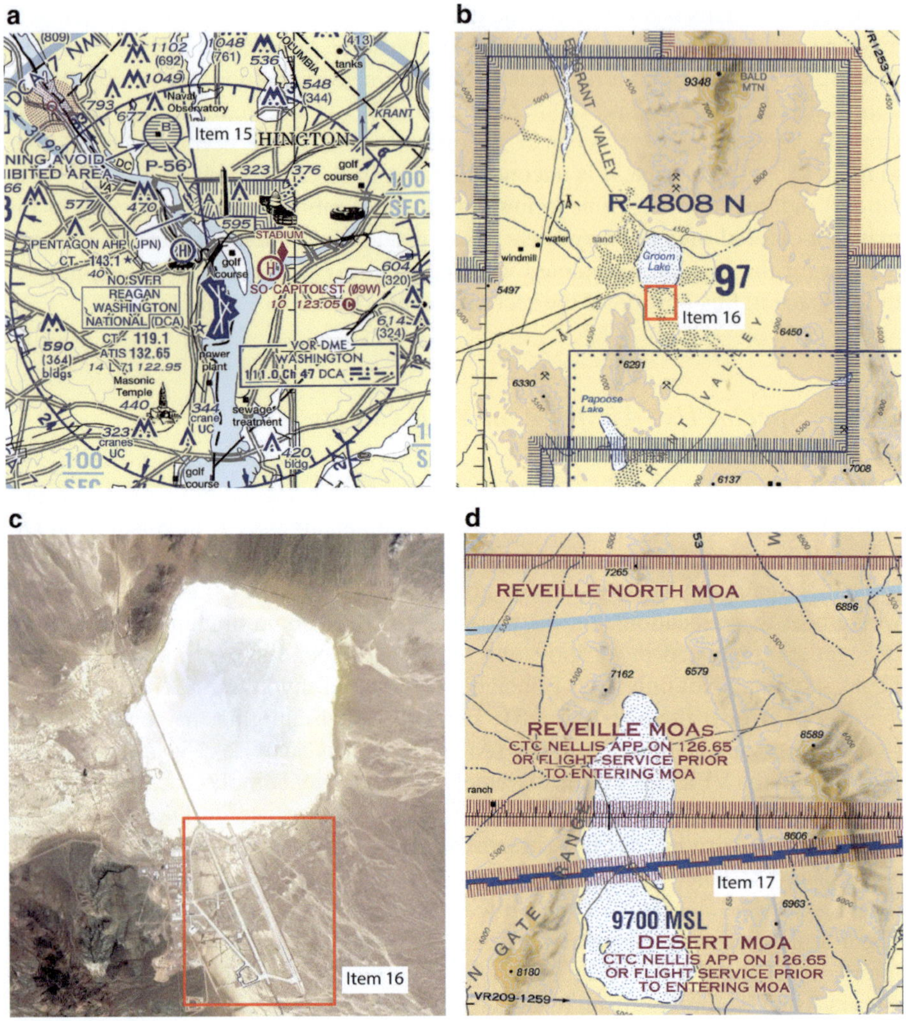

Fig. 3.10 (**a**) Prohibited areas in Washington, (**b**) Restricted area in Groom Lake, NV, (**c**) Area 51 runway, (**d**) Reveille military operating area

operators, this is the type of complex flying environment that would require significant pre-planning and attention to detail during operations. There is a lot going on in this part of Phoenix.

Finally, Fig. 3.11b highlights a warning area (W-174 C) off the southern coast of Key West, Florida. Again, warning areas are locations in coastal waters where the federal government does not have sole jurisdiction, but there may be high volumes of pilot training or unusual aerial activity. This is likely connected to Naval Air Station Key West, which is, among other things, a training facility for air-to-air combat fighter aircraft.

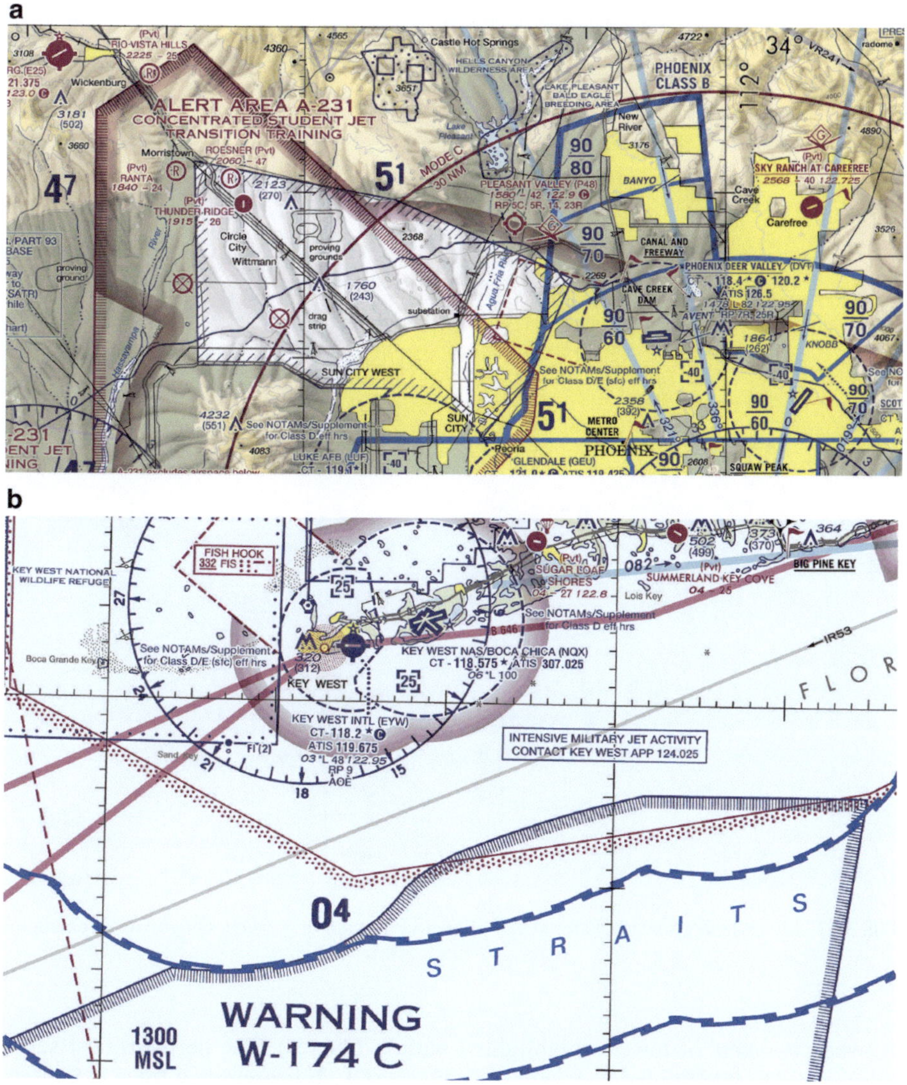

Fig. 3.11 (**a**) Alert area near Phoenix, AZ, (**b**) Warning area near Key West, FL

3.3.3 Miscellaneous

There are a handful of additional elements of the VFR sectional charts that users also need to be aware of. Figure 3.12 displays the VFR sectional for Eugene Oregon and its airport (EUG). Readers will notice a combination of Class Delta and Echo airspace in the area, with Echo starting at 700 ft. Also of interest are the array of local landmarks for the Eugene area. For example, consider [*Item18*], which is located

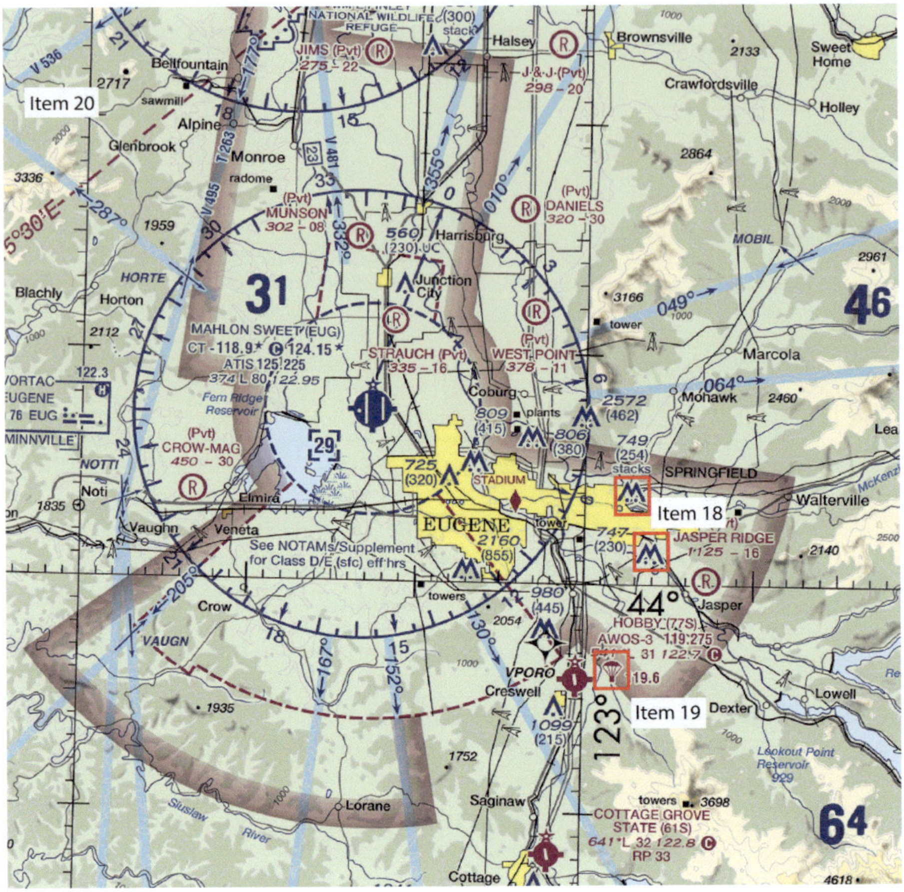

Fig. 3.12 Miscellaneous symbols for sUAS pilots to be aware of Eugene, Oregon as an example

between two sets of towers (highlighted with red boxes). The northernmost set is located at 749 ft MSL, but they extend to 254 ft AGL. It is important for sUAS pilots to recognize the AGL number as the most relevant for operations. These towers (which have some additional context associated with them) labeled as "stacks" are 254 ft high. This is well within the altitude that an sUAS would be operating (i.e., below 400 AGL) in the area. As a result, these stacks can (and should) be considered obstacles. The southern set could also be problematic because they are only 230 ft AGL. These kinds of obstacles become obvious once sUAS pilots are in the field. However, estimating the height of obstacles in the field is not something that humans are particularly good at doing. This is yet another way in which the VFR charts are helpful in planning missions. Consider [*Item19*] in Fig. 3.12. This represents an area where parachute jumping occurs with some frequency—an activity that would not be particularly sUAS friendly. [*Item20*] is located just below a light blue line on the chart, labeled V536. This is an example of a victor airway. The best analogy for victor airways is that they are the interstate highways of the sky. They are protected airspace for VFR aircraft, but can also be used by IFR to navigate between VOR stations. Operations along these airways extend from 1200 ft AGL to (but not including) 18,000 ft AGL.

3.4 Conclusion

The purpose of this chapter was to provide readers with a primer on the visual flight rules (VFR) charts from the Federal Aviation Administration (FAA). Understanding these charts is critically important for both mission planning and the safe operation of an sUAS. In particular, understanding the nuances of the national airspace system is essential for operating sUAS. Consider, briefly, the spate of drone incidents at Gatwick Airport (London) during December 2018 (BBC 2018). Several hundred flights were delayed, and tens of thousands of people were inconvenienced (or worse) because of unpermitted sUAS activity around the airport. Recent estimates suggest that the drone-related delays cost EasyJet nearly $20 million (Kollewe and Topham 2019). Although no arrests have been made at the time this chapter was written, this is the last thing the sUAS community needs. There is little doubt that sUAS incursions will continue to occur at airports. Do not be responsible for one of these. Know where you are flying at all times and fly responsibly.

References

Air Traffic Activity System [ATADS]. (2018). *Airport operations*. Retrieved June 14, 2019, from https://tinyurl.com/y7bvohv8

BBC. (2018). *Gatwick airport: Drones ground flights*. Retrieved from https://tinyurl.com/y9j34dnq

Federal Aviation Administration [FAA]. (2008). *Pilot's handbook of aeronautical knowledge. FAA-H-8083-25A*. Washington: United States Department of Transportation.

Federal Aviation Administration [FAA]. (2016). *Part 107 guidelines*. Retrieved from https://tinyurl.com/y8quosb4

Federal Aviation Administration [FAA]. (2018). *Pilot's Handbook of Aeronautical Knowledge*. Retrieved on July 26, 2019, from https://www.faa.gov/regulations_policies/handbooks_manuals/aviation/phak/

Grubesic, T. H., Fuellhart, K., Wei, F., & O'Connor, K. (2017). Regional perspectives on general aviation and reliever airports: A case study of the Phoenix metropolitan area. *Regional Science Policy & Practice, 9*(2), 101–120.

International Civil Aviation Organization [ICAO]. (2019) *Aeronautical charts*. Retrieved from https://tinyurl.com/yay2b6zh

Kollewe, J., & Topham, G. (2019). EasyJet says Gatwick drone chaos cost it £15m. *The Guardian*. Retrieved from https://tinyurl.com/y8ond9aa

Lumb, D. (2016). Disney can fly drones after unblocking its parks' no-fly zones. *Engadget*. Retrieved from https://tinyurl.com/op3nxsk

Marsh, A. (2008). Don't ask, don't tell: Area 51 gets airport identifier. *AOPA*. Retrieved July 25, 2018, from https://tinyurl.com/yc9jyobl

Roeder, T. (2018). Army's gray eagle drones will soon be ready for war. *The Gazette*. Retrieved from https://tinyurl.com/yb9jjquf

Terwilliger, B., Ison, D. C., Robbins, J., & Vincenzi, D. (2017). *Small unmanned aircraft systems guide: Exploring designs, operations, regulations, and economics*. Washington: Aviation Supplies & Academics.

Warnica, R. (2018). Janet, the mysterious airline that serves Area 51 is hiring. Applicants with top-secret clearance are preferred. *National Post*. Retrieved from https://tinyurl.com/yb7y5tdv

Chapter 4
Legal, Environmental, Operational, and Safety Challenges in Urban Areas

Abstract Small unmanned aerial systems (sUAS) are sparking a revolution in the way that information about the urban environment is collected. As the popularity and capability of sUAS continue to grow, the availability of high-resolution data will also surge. Advancements in power sources, global positioning systems, on-board sensors, and sUAS hardware means they can fly farther, faster, carry larger payloads, and capture imagery at an increasingly fine resolution. However, the rapid evolution of these technologies generates a variety of legal, environmental, operational, and safety challengers in the urban landscape. The purpose of this chapter is to detail these constraints on sUAS operation. This includes an overview of the operational and legal guidelines developed by the Federal Aviation Administration (FAA) and the Part 107 rules for operating sUAS, as well as the many challenges inherent to safe operation in different landscapes.

Keywords Legal · Federal Aviation Administration · Regulatory · Part 107 · Knowbeforeyoufly · Landscape · Obstacles · Hazards · Visual line of sight · Operations · Remote pilot in command · Limitations · Batteries · Safety · Geofences

4.1 Legal Challenges

There are a variety of legal issues centered on operating an sUAS in urban areas. These restrictions and constraints are constantly evolving (Marshall 2016). With the most recent release of the FAA Reauthorization Act (FAA 2018a), there is more clarity regarding where FAA jurisdiction stops and where state and local regulations begin for sUAS operations. This is an important advancement because of the ambiguity pertaining to sUAS operational guidelines within the 2012 FAA Modernization and Reform Act (FAA 2012). Specifically, many states and local governments were left questioning how to regulate sUAS activity. At the time, it was clear that the FAA was responsible for designating airspace, creating general operating and air traffic control rules, as well as the regulatory options for specialized aircraft (Raley 2017). However, the FAA did not address any regulations concerning privacy and trespass,

T. H. Grubesic, J. R. Nelson, *UAVs and Urban Spatial Analysis*,
https://doi.org/10.1007/978-3-030-35865-5_4

noting that this should be left with local governments. This fueled the uncertainty in sUAS regulation as local municipalities and government agencies attempted to fill these gaps without risking federal preemption. In the subsections below, we detail several approaches used to deal with sUAS use at the state and local level.

4.1.1 State and Local Regulatory Activity

One major challenge in developing regulations for sUAS at the local level is the possibility that federal regulations could preempt any efforts. For example, any state or local law that restricts where and/or how an sUAS can fly denotes sUAS operational limits within a city or above landmarks or creates laws requiring special permits, equipment, or training risks being preempted by federal law. Regardless of the preemption potential, cities across the United States began enacting ordinances aimed at limiting sUAS operation at about the same time sUAS use started to grow (circa 2013). For instance, beginning in 2014, some cities in Southern California (e.g., Beverly Hills and Hermosa Beach) created aggressive sUAS ordinances. Hermosa Beach, for example, made it a misdemeanor to photograph or record videos of any individuals within the city limit. They also require operators to register their sUAS with the city and purchase a permit for $145 (Hermosa 2016). Where the former is concerned, violators can be fined up to $1000 and face potential jail time. The latter represents an effort to limit the number of sUAS operating within the city but is also a clear violation of FAA regulations and risks federal preemption should it ever be challenged in court.

At first glance, local regulations for sUAS may seem difficult to impose, but there are a few options that local governments can pursue. First, the designation of where an sUAS can launch/land is within the scope of a municipality as it is linked to local zoning laws. Second, it is conceivable that a municipality could incorporate FAA regulations directly into local code. It is important to note that municipalities taking one (or both) of these approaches can then enforce these regulations through local law enforcement agencies. An ability to confiscate, levy a fine, or kindly ask operators to stop flying the sUAS (best case scenario) would then be within the purview of the city. Per the FAA (2018b):

> Laws traditionally related to state and local policy power—including land use, zoning, privacy, and law enforcement operations—are not subject to federal regulation…Cities and municipalities are not permitted to have their own rules or regulations governing the operation of aircraft.

In addition to these federal constraints, local ordinances (city level) must also contend with the potential for state-level preemption. According to the Association for Unmanned Vehicle Systems International (AUVSI 2018), the world's largest nonprofit organization for the advancement of unmanned systems, as of September 2019 there are 269 state-level ordinances aimed at sUAS operation that either have been introduced, passed, and currently being debated or were introduced and failed. Of those ordinances, 84 detail crimes, 26 detail law enforcement use, 30 are specific to weaponization of an sUAS, and 23 detail some aspect of privacy.[1] If nothing else, this is an indication of the magnitude of activity by state governments to monitor and control how sUAS are used. At the time this chapter was written, only seven states do not have some sort of regulation regarding sUAS use.

Other approaches to regulating sUAS include the insertion of additional language to existing privacy and harassment ordinances. For example, *New York Bill A4642* amends their existing civil rights law to make it unlawful to use an sUAS for unwarranted surveillance of a neighbor (Curran 2017). In Illinois, *House Bill 3906* amended the criminal code to make it unlawful for any person to operate an sUAS within the confines of another person's private property while recording video or images—or to invade another person's reasonable expectation of privacy (Moylan and Olsen 2017). As one might expect, these type of laws and regulations are *not* consistent as one moves from state to state or city to city. In fact, this patchwork of regulation presents significant challenges for sUAS operators at the local level. Fortunately, federal-level regulations for commercial sUAS operators did not change dramatically with the most recent FAA Reauthorization Act. As mentioned earlier, it actually clarified some of the confusion between federal, state, and local laws. These changes are detailed below.

4.1.2 Federal Changes to sUAS Regulation

The 2018 FAA Reauthorization Act (2018 Act) included an array of new regulations for sUAS, especially for users who consider themselves hobbyists. Prior to the 2018 Act, the FAA took a hands-off approach to recreational use. If the sUAS was under 55 lbs, operated within visual line of sight (VLOS), and did not fly at night, very few formal regulations existed. Furthermore, these were merely guidelines for recreational operators, not formal regulations. This was not the case for commercial operators, who were required to file for a Section 333 exemption and register their sUAS with the FAA (2012). The Section 333 requirement changed in 2016 with the issuance of the Part 107 small sUAS rule. This gave sUAS (under 55 lbs) operators a streamlined path toward legal operation. Under the new rule, commercial sUAS operators could take and pass the Part 107 test at a designated FAA testing center to establish themselves as a remote pilot in command (RPIC). Along with this was a set of operational and aircraft restrictions for commercial use, all of which are detailed on the FAA website (FAA 2016).

With the origination of Part 107 sUAS rules, there was some confusion about the distinction between commercial and recreational sUAS operators. For example, until the 2018 Act, recreational operators were not held accountable for detailed sUAS operational knowledge. Recent work suggests that this gap in training and knowledge for recreational operators was viewed as being particularly problematic.

[1] For a map of states that have sUAS regulation, see https://cqrcengage.com/auvsi/statelegmap.

Studies show that both experts and laypersons familiar with sUAS believe that recreational operators pose the largest risk to both public safety and privacy (Finn and Wright 2012). As a result, it is not surprising that the 2018 Act repealed Section 336 which contained the language used to exempt recreational operators from FAA regulation.

The removal of the Section 336 exemption for recreational operators means that all sUAS, regardless of their operator, are considered unmanned aircraft. Again, this designation eliminates the model aircraft exemption, which had historically placed recreational operators outside the purview of the FAA. To make things more confusing, the FAA does allow for the operation of recreational sUAS without formal certification if the following qualifications are met:

1. Flown strictly for recreational purposes.
2. Operated in accordance with the programming of a community-based organization's set of safety guidelines that are developed *in coordination with the FAA*.
3. Must stay out of the way of manned aircraft.
4. Must be flying in Class G airspace under 400 ft, or have authorization.
5. Must be registered and marked.
6. The operator must pass an aeronautical knowledge test.

There are a few interesting points to be made about the new recreational requirements. First, recreational sUAS operators need to be aware of the possibility of a required knowledge test. The FAA claims that it will not be identical to the Part 107 exam (which is relatively thorough), and it is still uncertain whether it will actually be implemented. At the time this chapter was written, there was no required test for recreational operators, yet it is still mentioned in Section 349 of the 2018 Act. Regardless, this requirement represents a big change to the existing rules considering that sUAS are marketed as an all-ages hobby platform. Second, the 2018 Act explicitly requires that hobbyists fly under 400 ft, where in the past, this altitude limit was only a guideline. Finally, the 2018 Act directs the FAA to review privacy concerns pertaining to sUAS operation. In the future, this may include a requirement for commercial operators to develop and maintain a written privacy policy which can then be made available to the public. It would be wise for readers to actively monitor these developments so as to avoid any unwanted financial penalties or, worse, a misdemeanor.

4.1.3 Spatiality of Legal Challenges

As detailed in Chap. 3, the FAA's Knowbeforeyoufly website (http://knowbeforeyoufly.org/) is an important resource for sUAS operators. It was founded by the two nonprofit advocate groups for unmanned vehicles, the AUVSI (mentioned previously) and the Academy of Model Aeronautics (AMA). In partnership with the FAA, the website serves as a resource for sUAS operators to plan and clear flights prior to launching the sUAS. Aside from Buzzy, the completely silly Knowbeforeyoufly mascot, the website includes a variety of resources to help

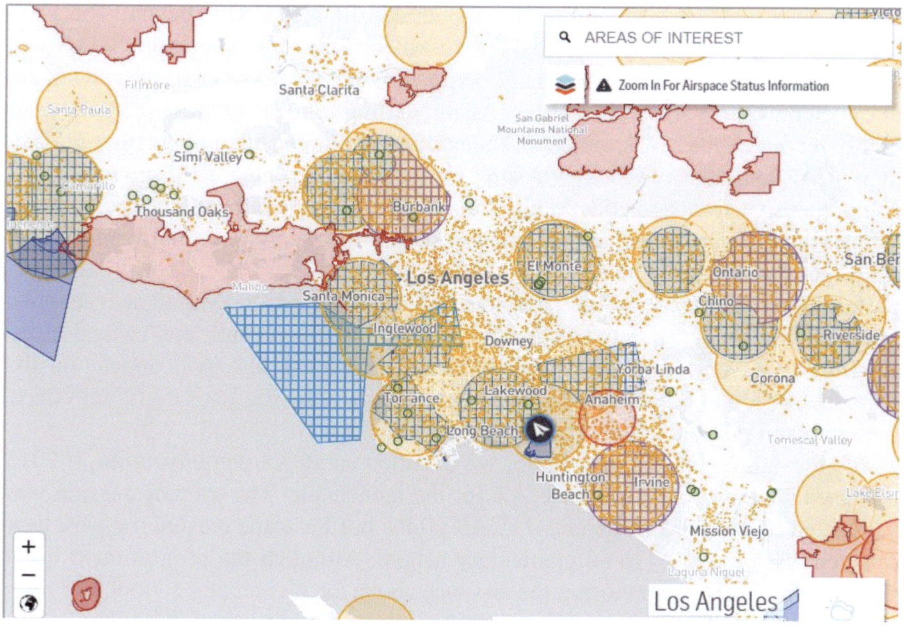

Fig. 4.1 An example of the Knowbeforeyoufly map indicating permanent flight restrictions, temporary restrictions, and other areas to be aware of while operating an sUAS (Federal Aviation Administration)

operators check rules and regulations and explore current flight restrictions using a web-based mapping application (Fig. 4.1). With the recent release of the Knowbeforeyoufly mobile application, operators can check real-time flying conditions (e.g., wind speed and direction, visibility, storm paths) and compare operation locales to the known controlled airspace boundaries.

While Knowbeforeyoufly certainly represents a positive step forward for making sUAS operations safer, it does not include any information about specific laws and ordinances at the city and state level. As mentioned previously, while it is true that most states take a fairly general approach to sUAS rules, primarily codifying FAA guidelines where necessary, others have modified the wording of their privacy and trespass rules to include sUAS operation. The AUVSI website (https://www.auvsi.org/) is a good reference for determining the existence of state-level regulations, even if the website is updated at slow intervals.

There are relatively few city ordinances that limit sUAS operations in ways considered stricter than the FAA rules, but they do exist. Sadly, there is no comprehensive resource that details these restrictions. At the time of writing, the best resource is made available by the Bard College Drone Center (https://tinyurl.com/y6eb435q). It details the nature of each restriction and provides the associated verbiage. What is unclear from the Bard College resource is how often this information is updated. For readers of this book, we recommend making a quick call to city offices to determine if there are any restrictions. This is much easier than dealing with local law enforcement and getting your sUAS impounded. If restrictions beyond the FAA regulations do exist, they are likely focused on locations where the sUAS can launch/land. Many cities (including Phoenix, AZ) have parks or open spaces that are designated as drone-friendly locales. Again, calling to identify these restrictions is highly recommended. Alternatively, many municipalities and counties that publish their sUAS guidelines to local websites. For example, Irvine, California provides a comprehensive webpage concerning the use of drones within the Irvine city limits (https://tinyurl.com/y34g3pc6). In short, a simple web search may suffice for many operators and operating locales.

Additional flight restrictions from the FAA include sUAS operation near schools, over crowds of people, state and federal prisons, sporting events, and critical infrastructure (e.g., power plants). Many of these restricted sites will appear on the Knowbeforeyoufly map, but others will not. Again, planning your flight with these limitations in mind is critical for a successful mission.

In an example of what *not* to do, we attended an sUAS demonstration in 2015 that used a local park in Phoenix, AZ for the launch site. The training session was conducted by certified commercial sUAS pilots, but for some reason, the park they selected was next door to an elementary school. Although the programmed flight was structured to avoid entering the airspace above the elementary school, the waypoint for the drone was mis-programmed to occur directly above the school. As a result, during the landing phase, it swooped over the school, decreased in altitude and eventually landed in the park. There were no mishaps, but if somebody had filed a report about the sUAS flying above the school, it could have ended with a revocation of the commercial license and/or a financial penalty.

A second interesting aspect of this anecdote is that flight planning software was not particularly good in 2015. Today (2019), not only is the planning software much improved, there are scores of internal safety features in both the software and hardware. In modern flight planning software, it is much easier to program launch/landing locations, home waypoints, geofences, and the boundaries of the survey area. On the rare occasion that the sUAS loses connection with the telemetry unit and/or control hub, suffers from a sudden drop in battery power, or needs to make an emergency landing, the flight planning software allows pilots to designate a safe space for these activities to take place. Of course, there is an increased risk of violating locally restricted airspace when this happens, but careful planning of the mission parameters can go a long way to ensuring the safe execution of emergency protocols.

4.2 Environmental Challenges

Flying an sUAS in a dense urban or suburban environment poses a unique set of challenges to pilots. Not only are there more people and vehicles, but there are more structures, radio interference, and airspace restrictions to be concerned with. In the subsections below, we detail a range of problems and solutions for dealing with complex urban and suburban landscapes.

4.2.1 Local Landscape

The local landscape is an important consideration when flying an sUAS. Broadly defined, the local landscape consists of all the built structures and infrastructure in a working location. This includes office buildings, single-family residential homes, multifamily apartment buildings, streets, dams, radio towers, electric lines and poles, etc. It also includes local environmental, ecological, and physiographic features, such as parks, trees, rivers, mountains, and canyons. In sum, all of these structures and associated local features must be considered when planning an sUAS mission. For example, consider the basic challenge of working proximal to several tall buildings. Not only does the pilot need to be aware of the height of the buildings, the sUAS mission needs to be programmed to account for these heights and to maintain a safe operating distance from the buildings. While modern flight planning software does a decent job of incorporating elevation information from digital basemaps, elevation tiles rarely include information on built structures. This can be problematic because it is difficult for humans to estimate heights of structures and/or landscape features without a lot of practice. Thus, while some flight systems are starting to incorporate collision detection, these features remain under development (Gageik et al. 2015; Bareiss et al. 2017).

At this juncture, it is important to note that increasing levels of autonomy in sUAS systems require more environmental awareness and pre-flight planning for their operators. This may seem counterintuitive, but reconsider the challenges presented by the local landscape. Pilots using systems with pre-programmed takeoff, landing, and flights need to be cautious when it comes to the rate that the sUAS climbs and descends, as well as *where* this climb/descent takes place. This is not easy to control for many systems. In another example of what *not* to do, an operator recently crashed an sUAS into a construction crane at a building site in Kent, England (Holligan 2018). Specifically, the sUAS was launched by the operator with the goal of executing a preplanned mission to inspect a building that was under construction. The mission, planned and programmed a few weeks earlier, included four waypoints that the sUAS would autonomously fly to, all of which were at an altitude of 400 ft (to avoid buildings and construction equipment in the area). However, while climbing to the first waypoint, the sUAS crashed into a newly erected crane—one that was not present (or considered) when the mission was initially planned. The incident was reported to Air Accidents Investigation Branch (AAIB), where the pilot noted that in the future, he would "plan such flights to ascend vertically to 400 ft shortly after take-off, in a safe corridor, and would avoid cranes" (Holligan 2018). If nothing else, this underscores that awareness of the ambient environment and local landscape is crucial to sUAS pilots and spotters. The landscape is dynamic, changing hour-to-hour, day-to-day, and week-to-week.

The undoing of the sUAS mission in Kent was partly attributed to a lack of visual perspective in/around the operating area and the new crane, but also a somewhat lazy, angular ascent of the quadcopter sUAS. Obviously, quadcopters can ascend directly upward, which makes the Kent incident more embarrassing. However,

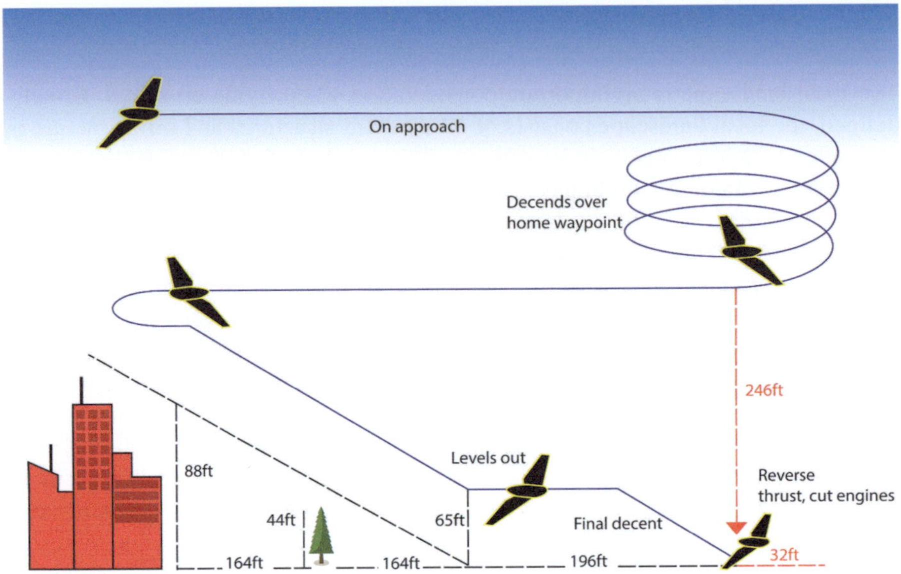

Fig. 4.2 eBee landing behavior

many fixed wing sUAS are particularly susceptible to these types of problems as their takeoff and landing occur at an angle (Fig. 4.2).

Consider the eBee platform, which is one of the more popular fixed-wing sUAS used for mapping and surveying applications. For readers with any experience operating an eBee, the following information will not be surprising. First, landing the eBee can be (but not always) a gut-wrenching experience. After completing image collection and returning to the home waypoint, the eBee is programmed to head away from the landing zone to make its final approach. Immediately after its final turn, it begins to descend rapidly. When it is located 500 ft away from the landing zone, it begins to dive. At 230 ft out from the landing area, the eBee is at an altitude of 60 ft and when 100 ft away from the landing zone, the eBee reverse thrusts and then cuts the motor for a glide landing. To be clear, after several hundred flights, we are never completely certain where the eBee will touch down. Depending on ambient conditions (especially wind), we have experienced some drift in landing locations. Operationally, this means that "close enough" landings will typically suffice when the site is relatively free from obstructions. However, when flying in urban spaces where the operational space is tight, we are forced to lean heavily on previous flight experience(s) when evaluating landing sectors during flight planning. In good flying conditions, we have "threaded the needle" between trees, cacti, and other obstructions on many occasions. That said, in poorer flight conditions, moderate winds have pushed the eBee into Palo Verde trees, ocotillo (which may be the scariest/thorniest plant in the Sonoran Desert), and other local flora. In practice, we try to maintain at least a 100-ft buffer on both sides of the landing zone and at least

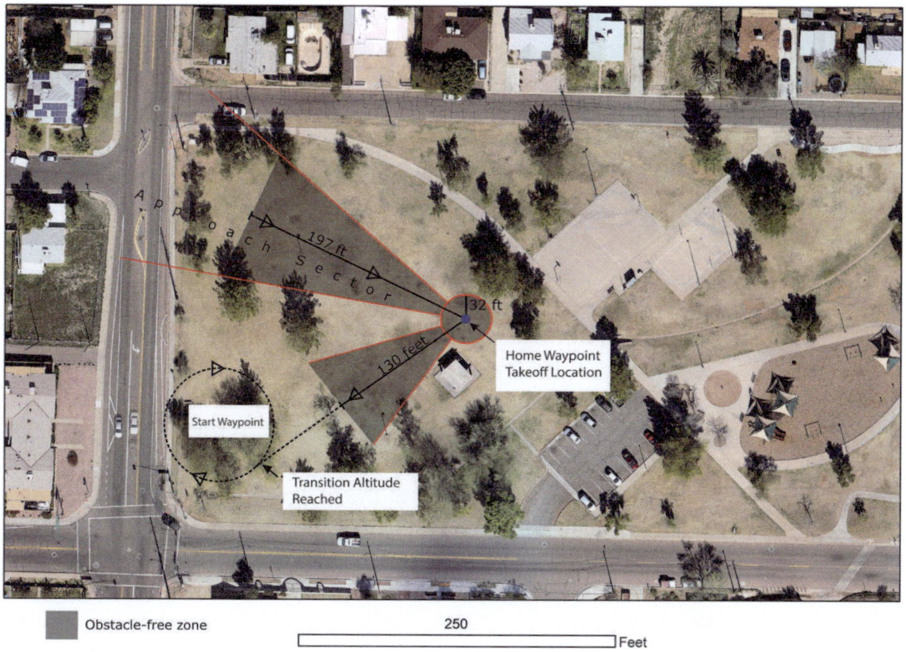

Fig. 4.3 Landing considerations in a complex urban environment

a 500-ft buffer in front of the landing area to account for any sudden fluctuations in the wind.

Figure 4.3 displays a landing and takeoff schematic for the eBee, superimposed over one of our study areas where "threading the needle" was particularly important. Of note is the position from where the eBee starts its flight and the location it performs its initial vertical climb. This requires a fairly large amount of open space (130 ft cone for takeoff and 197 ft cone for landing) with no obstructions for a "drama-free" takeoff. Figure 4.3 also highlights the uncertain nature of the approach sector. In this situation, there was only one obstacle-free approach sector, and we really needed to leverage this zone for the landing. Any deviation would put the eBee into a tree (bad) or nearby powerlines (worse). Keep in mind, the eBee is programmed to make its landing approach anywhere in that sector. For approaches on the boundaries of the sector, a change in wind direction and/or speed can easily push the eBee outside of the planned approach. Similarly, a strong tail wind can cause the eBee to overshoot its home waypoint (something we have experienced a number of times). For the mission flown in Fig. 4.3 overshooting the home waypoint was much better than being blown off course. In this instance, distance calculations required precision, but we also hedged our bets by flying on a day with limited wind.

In some of our other field locations, the local elevation shifts rather dramatically. For the Wingtra sUAS platform, changes in elevation are considered by the flight planning software and relayed to the drone prior to the mission. This is also true when using the Emotion flight planning software by Sensefly for the eBee. As detailed in Chap. 3, flight ceilings are somewhat dynamic because they are linked to above ground level (AGL) elevation measurements. For example, if the elevation within the study area changes by 100 or more feet and the flight altitude is set at 400 ft AGL, some parts of the flight could flown at 500 ft above mean sea level (MSL) or higher. Again, we want to emphasize these nuances to readers—which are critical for flight planning and operations. Know the difference between AGL and MSL and plan accordingly. More importantly, if you are operating without sophisticated flight planning software, it is imperative that you understand the local landscape and its quirks when conducting and/or planning a mission.

Finally, there are some additional considerations pertaining to the weather that must be considered when operating an sUAS in complex urban environments. As noted previously, our experience with the lightweight eBee platform has provided a variety of reality checks on the impacts of weather (i.e., wind) for takeoff and landing. However, wind can also disrupt the data collection portion of the flight, pushing the sUAS off-kilter and impacting the quality of the images. Although this typically amounts to nothing more than a blurred image, as the corpus of blurred images grows, it gradually degrades the overall quality of any postprocessed product created (e.g., digital orthophotograph). While some systems can account for changes in wind speed and direction during the flight, there are limitations in the amount of wind that an sUAS can tolerate. In some instances, wind can cause operators to lose control of their system. For example, high winds recently pushed an sUAS into the Barclays building in New York (Moores et al. 2018). High winds can also blow the sUAS beyond radio contact. For some systems, this means that the sUAS will likely crash (somewhere, or into somebody) and/or will be permanently lost (especially if this happens beyond VLOS). For those equipped with additional safety features, loss of radio contact prompts a "return to home" function in the sUAS, but the strong winds could prevent this return if battery power is low. In sum, situational awareness is critical to the safe operation of the sUAS, and this includes a thorough understanding of how weather impacts flight safety and longevity.

4.2.2 Local Air Traffic

When operating an sUAS in an urban environment, the likelihood of interaction with another aircraft increases. Obviously, this should be avoided at all costs and if the operator is following the FAA rules, this is rarely a problem. That said, not all aircraft operators adhere to the FAA rules, and we are not referring to sUAS pilots. For example, while helicopters should be flying at (or above) 500 ft, they are permitted to fly below that floor if "the operation is conducted without hazard to persons or property on the surface" (FAA 2010). In urban areas like Phoenix, Arizona, law enforcement helicopters fly below the floor somewhat regularly. If this situation does arise, right-of-way must be yielded to the helicopter. In the FAA commercial drone rules, Section 107.37 includes the rules for operation near other aircraft as well as right-of-way rules (FAA 2018d):

> Each small unmanned aircraft must yield the right-of-way to all aircraft, airborne vehicles, and launch reentry vehicles. Yielding the right-of-way means that the small unmanned aircraft must give way to the aircraft of vehicle and may not pass over, under, or ahead of it unless well clear.

For sUAS operators, this means that if any helicopter is in the area, they must abort their mission as quickly as possible. Autonomous systems, like the Wingtra or eBee, have a return-to-home function which aborts the current mission and immediately starts the decent and landing procedures. However, if a crash looks to be imminent, systems like the Wingtra have two other options. The first is a manual takeover of the flight which would allow the operator to begin a quick decent down to avoid an accident. The second is a fail-safe immediate abort which cuts power to the motors. The latter is not ideal, but it is better than a mid-air collision with a manned aircraft.

One way to judge the likelihood of having an interaction with a helicopter is by the presence of flight infrastructure in the local area, such as helipads. The Knowbeforeyoufly application and VFR charts have the locations of the heliports throughout the United States. As detailed previously, these resources must be consulted during the mission planning process.

4.3 Operational Challenges

We classify *operational challenges* for sUAS as those that would impede the use of a drone for routine data gathering missions. In the previous section, we detailed several weather-related challenges that stress the operational limits of a platform. These environmental factors place limitations on the duration of the flight and the quality of the images and can prompt the cancellation of a flight when conditions are severe. However, these factors differentially impact sUAS because each platform is unique. Yet, there are additional operational challenges that are embedded within a given system. For example, some systems restrict maximum altitudes, flight speeds, and the takeoff/landing routine. Autonomous systems will often have a cruising speed that has been optimized for precise data collection which cannot be changed. In addition, there are legal limitations that can turn into operational challenges, including the requirements that the remote pilot in command operates within the rules laid out by the FAA.

4.3.1 Administrative

If you are reading this book, it is likely that you have already dealt with the biggest administrative hurdle for operating an sUAS, which is taking and passing the Part 107 aeronautical knowledge exam (FAA 2018a). Passing this exam allows one to become a remote pilot in command (RPIC). An RPIC must be present during any sUAS mission. Part 107 was established in 2016 to speed up the certification and exemption process for commercial sUAS operators. Part 107 applies to any sUAS which is 55 lbs or under and represents a follow-up to an early process outlined in the 2012 FAA reauthorization act requiring that all commercial operators apply for a Section 333 exemption. This was an arduous process that was quickly overwhelmed by the massive number of requests being made (Marshall 2016).

Part 107 details many operational limitations with respect to the environment (FAA 2016). First, flights can only occur during daylight or twilight hours. This includes 30 min prior to and after the official sunrise and sunset times of a location. Second, local visibility must be *at least* three miles. Both of these operational requirements pertain to maintaining VLOS with the sUAS. Another important operational limitation prohibits the operation of sUAS above individuals not involved with the flight. This typically means that the sUAS can only fly above the RPIC and the spotter. However, if bystanders are under/within a protected structure or a vehicle, it is possible to operate an sUAS above them. The purpose of this regulation is to limit drones from flying above crowds at large sporting events, festivals, or protests. It is important to note that a waiver to operate above people can be granted by the FAA, after submission through the "Drone Zone" website (https://faadronezone.faa.gov), but they are extremely difficult to obtain. Details and associated information for previously granted waivers are also available from the site.

4.3.2 Aircraft Limitations

sUAS platforms are not equal. They are built for different uses and/or purposes, and some of this is reflected in the limitations of the sUAS components. Many platforms have significant limitations, including restrictions on ambient operating temperatures, wind resistance, and the like. For example, the Wingtra is designed to operate in temperatures from −4 to 122 ° F, while the eBee can operate between −15 and 122 ° F. However, these maximum operating temperatures are just that the *max* temperatures. In Phoenix, Arizona, summer temperatures can routinely reach 110 °F by mid-day and occasionally reach 120 °F or hotter. This is extreme heat. Even the most robust sUAS platforms can be damaged when operating in these conditions. Heat is hazardous to the optics, batteries, motors, and the computing equipment (e.g., tablet) used for flying the sUAS. Consider, for example, starting one's car in extremely low temperatures. The battery voltage is significantly reduced (Wang 2017; Zhang et al. 2006), making it more difficult for the engine to fire. Higher temperatures can lead to the accelerated corrosion of batteries, irreversibly damaging its internal structure and reducing its life. Both heat and cold can cause in-flight failure, aborted flights, and/or loss-of-control events.

As alluded to above, platform operating constraints vary. Table 4.1 provides more concrete details of these variations for several of the more popular sUAS. For example, the DJI Mavic Pro and DJI Phantom 4, two of the most popular sUAS systems, have max operating temperatures of 32–104 °F (DJI 2019). This is not suitable for most mid-summer days in Phoenix or winter days in the Midwest or

Table 4.1 A subset of different UAV systems and a few of their important operating limitations

Model	Type	Max low temperature (°F)	Max high temperature (°F)	Max speed (m/s)	Flight time (max, no wind)	Max wind resistance (cruise, mph)	Max altitude (ASL, ft)
Wingtra	Fixed	−4	122	16	55	28	9800
eBee	Fixed	−4	122	25	50	28	17,060
Quest DATAhawk	Fixed	−4	122	26	45	35	3000
Parrot Disco	Fixed	32	104	22	45	27	500 (restricted)
DJI Mavic Pro Platinum	Quad	32	104	18	30	Operator limited	16,404
DJI Phantom	Quad	32	104	20	28	22	19,685
Parrot Anafi	Quad	14	104	15	25	31	14,763
Autel Robotics EVO	Quad	32	104	20	30	Operator limited	2600

New England. In fact, most copter-model systems have the same max operating temperatures as the DJIs—which are significantly more limited than many fixed wing systems (Table 4.1).

As detailed previously, wind is also a significant operational challenge. Maximum wind resistance varies from platform to platform. Generally speaking, the lighter the sUAS, the lower the wind resistance. Much of what factors into the maximum wind resistance recommendation pertains to the ability of an sUAS to return home successfully. For example, a low battery combined with a strong headwind can result in a mid-air loss of power and a crash. Continual compensation for wind will also lead to modest variations in image overlap and can lead to blurred imagery, more generally. In every respect, the sUAS is working much harder in windy conditions.

Other operational limitations are independent of the weather conditions and based on the design of the system. Maximum speed varies across all systems, as does maximum altitude and flight endurance (with no wind). One major operational challenge faced by all sUAS is the balancing of payload weight and flight time. Generally speaking, more weight equals less flying time, but this is highly dependent on battery sizes and the overall flight efficiency of a platform. Regardless, size and weight can restrict the amount of data that can be collected in a single flight. For example, consider the eBee, which weighs about 1.5 lbs with the camera and battery. The eBee has a maximum flight time of approximately 50 min when using the Venom 3S 2200 mAh 11.1 V LiPo battery.[2]

Conversely, the Wingtra has a reported flight time of 55 min, but its weight with batteries and camera is nearly 7 times as much as the eBee, at just under 10 lbs (Wingtra 2018). A heavier aircraft means more stability in the air and larger payloads (bigger cameras), but that comes at the expense of flight time if the size of the batteries (and weight) are not increased as well. In this instance, the Wingtra is powered by *two* 4S3P lithium-ion batteries—a much larger battery payload when compared to the eBee, but with significantly more energy to power the larger platform.

Finally, one of the most important operational challenges for urban remote sensing is image resolution and ground sampling distance (GSD). GSD is the size of the pixel in the image and relates to the smallest object that the sUAS is able to resolve (Jensen and Lulla 1987). The GSD depends on the quality of the camera (megapixels or MP) and altitude from which data is being collected. The formal calculation is:

$$GSD_h = \frac{\text{Flight Height} * \text{Sensor Height}}{\text{Focal Height} * \text{Image Height}} \tag{4.1}$$

As the altitude increases, the GSD gets larger. This translates to a degraded ability to resolve small objects in the imagery. However, at the same time, when altitude increases, so do image footprints, allowing for more geographic coverage (Fig. 4.4). Thus, while higher resolution images (small GSD) are possible by flying at a lower altitude, the image footprints are smaller, which yields longer flight times to capture the entire area of interest. This is an important and interesting tradeoff for sUAS operations. In an urban environment, time (and time-sensitive data collection) can be a critical factor because the window for gathering data may be small. In this case, operators may need to sacrifice image resolution for larger image footprints in order to achieve a decreased flight time. As detailed above, another limiting factor is battery life. Data collection at lower altitudes may require more than one flight to capture imagery for the entire region because of the operational limits of the onboard batteries. For platforms like the Wingtra, there is an ability to conduct a mid-mission battery swap, where the sUAS would resume data collection at the point it suspended operations for the battery change. One way to ameliorate these limitations is to use a higher resolution camera that can obtain smaller GSD at higher altitudes. Of course, many platforms have limited choices

[2] Battery nomenclature can be really confusing for the uninitiated. In this case, "3S" means a three cell square (or rectangular) battery. 2200mAh refers to the amount of charge inside the battery and the "mAh" stands for milliamps per hour. Remember, it takes energy to transfer energy. So, the real capacity of a power cell is about 63% of the reported number. In this case, a 2200mAh battery has a real capacity of 1380mAH because 820mAh is lost in the transfer process. "11.1 V" is a reference to voltage. In this example, the 3S arrangement equates to 3.70 V∗3 cells or 11.1 V. Finally, "LiPo" refers to lithium polymer cells which are used when weight is a critical consideration in the device that needs to be powered. Simply put, LiPo batteries provide more power at a lighter weight than lithium-ion or lithium-metal batteries.

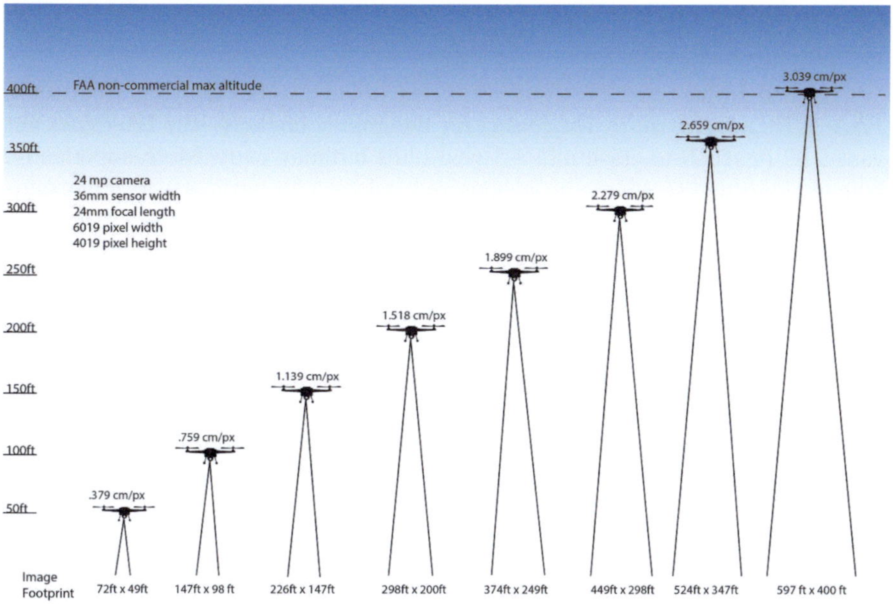

Fig. 4.4 Relationship between flight altitude, camera focal length, and ground sampling distance

for onboard cameras. As of March 2019, the eBee platform has seven compatible cameras while the Wingtra has four. Payload limitations and the size of the camera bay may restrict camera availability.

4.4 Safety Challenges

In some ways, the safety challenges detailed in this subsection represent a compendium of the material already covered in this chapter. As one might expect, a quick Internet search on "drone crash" yields several pages of recent incidents. Readers need to remember that it is always the operator's responsibility to ensure that the sUAS is not being used in way that could inflict damage to the flight team, bystanders, property, or other aerial vehicles. Operators who have passed the Part 107 test should be intimately aware of the operational restrictions and safety hazards associated with sUAS. If anything, the material in this chapter emphasizes that sUAS operations in urban environments present a unique suite of safety challenges, particularly when compared to rural or remote areas, where property and people are more diffuse. Taking steps to address the challenges detailed here will drastically reduce the likelihood of negative outcomes when operating sUAS. Furthermore, advances in sUAS technology are helping to ensure that drones are operated safely by both commercial and recreational operators.

4.4.1 Flying Over People

Part 107 and the associated FAA rules are clear when it comes to flying over people (FAA 2018e). *Don't do it*. The reason for this rule is obvious—the FAA does not want a 30 lb sUAS to crash into a 5-year old's birthday party. More importantly, there are fears that sUAS can be weaponized and used for terror attacks. For example, in August 2018, military-grade explosives were attached to two commercial drones in an attempt to assassinate the president of Venezuela, Nicholas Maduro (Walsh et al. 2019). Although Maduro was not injured, seven members of the Venezuelan National Guard were wounded (Walsh et al. 2019). This grim application of sUAS technology is difficult to prevent, but as sUAS systems become smarter and more adaptive, it is highly likely that sensing technology will be developed to help avoid people and crowds. For now, systems equipped with automatic return-to-home features (e.g., if radio connections are lost, or battery power is running low) can help prevent unintended interactions with individual bystanders or larger groups of people.

It is also important to remember that operating in urban environments require a *very* good sense of local operational context and the conditions of where the mission is taking place. For our own research, we have spent many hours in/around the locations where our sUAS operates, some of these areas are more sparsely populated than others. Regardless of the local housing and population density, we conduct extensive surveys of the locations prior to any mission to document the activity level during different parts of the day. One notable advantage that is afforded by operating in Phoenix is the weather. By late morning in the fall, spring, and summer months, temperatures are approaching 90 °F. Although uncomfortable, these temperatures are well within the safe operating limits of most sUAS, and it drastically reduces our chances of flying missions over bystanders. As a result, we typically fly in the late morning. Still, there have been several occasions when we have delayed the flights due to the presence of bystanders in the area. In every scenario, waiting for people to vacate is better than dealing with the local law enforcement (following a call from a bystander) or reporting an incident to the FAA.

4.4.2 Flying Near Buildings

Flying near buildings or large structures can be a complex endeavor. Again, this is a major concern for public safety. In September 2019, 10 kamikaze-style drones with explosive warheads were used to disrupt oil production in Saudi Arabia (Dormehl 2019). The drones flew directly into the oil production structures. Needless to say, when flying an sUAS near any building, there are a number of operational and safety regulations that must be adhered to. For example, if the operator of an sUAS is conducting flights in a downtown area with buildings that are 300–400 ft tall, the sUAS can legally fly above the 400-ft ceiling as long as they are within 400 ft of the structure (FAA 2018c). In other words, if the building is 400 ft tall, the sUAS can legally fly at 800 ft. Please keep in mind that conducting these types of missions is very complex and would likely require some type of local authorization from air traffic control, especially if the mission is proximal to an airport or helipad or is within a local airport approach vector. That said, these types of operations *can* be conducted if the sUAS is operating at a safe distance from the building during the data collection effort. The usual caveats for safety apply. Buildings, especially clusters of buildings in central business districts (CBDs), often generate the Venturi effect where the pressure difference in the gaps between buildings produce pressure differentials that increase wind velocities (Blocken et al. 2008). As detailed earlier, a random gust of wind can throw a UAV off course and send it careening into a building, risking property damage and harm to individuals below. Because of this, some cities have enacted municipal codes to ensure that these types of operational problems will not happen. For example, New York City administrative code *10-126 C* prohibits piloting aircraft (of any type) in the city that are not related to safety.

4.4.3 Geofences

One interesting development associated with sUAS operations and safety is the use of a geofence. For RPICs, a geofence typically corresponds to an invisible barrier, operationalized by distance, that represents how far away an sUAS can operate. Figure 4.5 highlights an example of a geofence in the Wingtra flight software. If the geofence is pierced, the sUAS automatically returns home or to a designated waypoint. This type of functionality is an excellent way to ensure that operators do not

Fig. 4.5 Wingtra pilot geofence function

lose the sUAS in a complex operating environment. Again, this is less of a concern in rural areas. However, one can also use geofences to prohibit sUAS from entering specific portions of the airspace, regardless of urban or rural status. DJI is a front-runner in this area and has implemented this capability in their flight software. DJI calls this the Geospatial Environment Online (GEO), which is built into the flight software of their systems (DJI 2018). It allows for the incorporation of real-time flight bans on the use of sUAS in certain areas. For example, in the United States, sUAS are being used to deliver contraband (e.g., drugs, cell phones) to prisoners (Schmidt 2015). The implementation of a geofence is one way to combat this activity. Geofences can be implemented for other types of critical infrastructure and key facilities as well, including power plants, airfields, and stadiums. Geofences can also be used to limit altitude, avoid collisions, and ensure that individuals are properly apprised of local sUAS restrictions before flying (DJI 2018).

4.5 Conclusion

Numerous challenges exist for the operation of sUAS, some of which are exacerbated by flying in urban environments. Although this chapter detailed many of these issues, it is by no means exhaustive. Urban environments are incredibly diverse. As a result, the challenges in Phoenix, Arizona may not directly overlap with those found in Portland, Oregon or Philadelphia, Pennsylvania. Regardless of where one is operating an sUAS, the best advice we can provide readers is to maintain a keen sense of awareness during all phases of the operation, but also of the local environment. Take the necessary steps to follow the FAA operating rules and any local regulations that may be in place. This will reduce the odds of a loss-of-control event and improve the overall safety of sUAS operations. As part of this effort, we recommend that the mission team conduct multiple scouting trips to mission locations before the sUAS is launched. This is not always possible, but it can help provide important environmental and operational context prior to the mission. Developing a deeper understanding of the local dynamics and taking steps to address any operational or safety concerns will help operators avoid costly mistakes.

References

AUVSI. 2018. 2018 state legislative map. Arlington: The Association for Unmanned Aerial Vehicles International. Retrieved June 12, 2019, from http://cqrcengage.com/auvsi/statelegmap

Bareiss, D., Bourne, J. R., & Leang, K. K. (2017). On-board model-based automatic collision avoidance: Application in remotely-piloted unmanned aerial vehicles. *Autonomous Robots, 41*(7), 1539–1554.

Blocken, B., Moonen, P., Stathopoulos, T., & Carmeliet, J. (2008). Numerical study on the existence of the venturi effect in passages between perpendicular buildings. *Journal of engineering mechanics, 134*(12), 1021–1028.

Curran, A. (2017). Civil Rights Law. A4642. 2015-2016 New York Legislative Session.
DJI. (2018). DJI refines geofencing to enhance airport safety, clarify restrictions. *Newsroom News*. Retrieved February 12, 2019, from https://tinyurl.com/y5hlam6k
DJI. (2019). *Mavic Pro Platinum Specs*. Retrieved from https://www.dji.com/mavic-pro-platinum/info
Dormehl, L. (2019). The drones used in the Saudi Arabia oil attack were not 'off the shelf'. *Digital Trends*. Retrieved July 15, 2019, from https://tinyurl.com/y6fflbcl
FAA. (2010). General operating and flight rules. Washington, D.C.: Federal Aviation Administration. Retrieved from https://www.law.cornell.edu/cfr/text/14/91.119
FAA. (2012). Modernization and Reform Act of 2012. H.R.658. 112th Congress.
FAA. (2016). Summary of small unmanned aircraft rule (part 107). *FAA News*. Retrieved June 21, 2016, from https://www.faa.gov/uas/media/Part_107_Summary.pdf
FAA. (2018a). FAA Reauthorization Act of 2018. H.R.302. 115th Congress
FAA. (2018b). FAA statement–federal vs. local drone authority. Washington, D.C.: FAA. Retrieved from https://www.faa.gov/news/press_releases/news_story.cfm?newsId=22938
FAA. (2018c). Fact sheet—Small unmanned aircraft regulations (part 107). Washington, D.C.: FAA. Retrieved from https://www.faa.gov/news/fact_sheets/news_story.cfm?newsId=22615
FAA. (2018d). Operation near aircraft; right-of-way rules. Washington, D.C.: Federal Aviation Administration. Retrieved from https://www.law.cornell.edu/cfr/text/14/107.37
FAA. (2018e). *Operation over human beings*. Washington, D.C.: Federal Aviation Administration.
Finn, R. L., & Wright, D. (2012). Unmanned aircraft systems: Surveillance, ethics and privacy in civil applications. *Computer Law & Security Review, 28*(2), 184–194. https://doi.org/10.1016/j.clsr.2012.01.005.
Gageik, N., Benz, P., & Montenegro, S. (2015). Obstacle detection and collision avoidance for a UAV with complementary low-cost sensors. *IEEE Access, 3*, 599–609.
Hermosa. 2016. Drone, unmanned aircraft and model aircraft. Hermosa Beach: City Council. Retrieved from https://tinyurl.com/y2nadw9c
Holligan, A. (2018). Drone hit newly erected crane during kent site survey. *British Broadcasting Channel*. Retrieved March 12, 2019, from https://www.bbc.com/news/technology-42718824
Jensen, J. R., & Lulla, K. (1987). *Introductory digital image processing: A remote sensing perspective* (4th ed.). London: Pearson.
Marshall, D. (2016). What a long strange trip It's been': A journey through the FAA's drone policies and regulations. *DePaul Law Review, 65*(1), 6.
Moores, T., Feuerherd, B., & Feis, A. (2018). Drone pilot arrested after crashing into midtown Bank. *New York Post*. Retrieved November 18, 2018, from https://tinyurl.com/y6k5n3gs
Moylan, M., & Olsen, D. (2017). Criminal Code Drone Privacy. HB3906. 100th Illinois General Assembly.
Raley, C. (2017). FAA UAS symposium. In *Local and state UAS enforcement authorities* (Vol. 40). Reston: Federal Aviation Administration. Retrieved from https://tinyurl.com/y3bdsqz2
Schmidt, M. (2015). Airmail via drones is vexing for prisons. *New York Times*. Retrieved April 22, 2015, from https://tinyurl.com/y4hgxj58
Walsh, N., Gallon, N., Perez, E., Castrillon, D., & Arvanitidis, B. (2019). Inside the august plot to kill Maduro with drones. *CNN*. Retrieved from https://tinyurl.com/y3ftw4tl
Wang, K. (2017). Study on low temperature performance of Li Ion battery. *Open Access Library Journal, 4*(11), 1.
Wingtra. (2018). *WingtraOne technical specifications*. Retrieved from https://tinyurl.com/y46lcxag
Zhang, S. S., Xu, K., & Jow, T. R. (2006). Charge and discharge characteristics of a commercial LiCoO2-based 18650 Li-Ion battery. *Journal of Power Sources, 160*(2), 1403–1409.

Chapter 5
Small Unmanned Aerial Systems and Privacy

Abstract A significant advantage of sUAS technology is its ability to collect high-resolution data at a large scale, including individual streets, yards, and houses. Recent advances in sUAS technology, including decreasing size, increased functionality, and improved affordability, mean that both researchers and practitioners can collect these data quickly and without a major coordination of people and resources. Given the relative ease of such efforts and the resolution of data that can be acquired, there are increasing concerns that both recreational and professional operators can collect information from locations that might be considered private. For some bystanders, private space may be defined by porches and decks with privacy screens. For others, a backyard fence or wall may represent private space. Whatever the definition, privacy is central to the debate on the fair and equitable use of sUAS technology in urban areas. Moreover, it is one of the major factors in local decisions to regulate *where* and *how* sUAS are used. The purpose of this chapter is to discuss the concept of privacy in relation to sUAS technology and detail the pragmatic implications of privacy-related restrictions for drone operation. This chapter concludes with a brief overview of sUAS ethics and the increased need for local awareness and cultural sensitivity when operating sUAS.

Keywords Privacy · Ethics · Closed circuit television · Private property · Recreation · Hellfire

5.1 A Primer on Privacy

Individual privacy rights have been a topic of public discourse and academic research for over a century. This includes a focus on the evolution and use of surveillance technology (Rachels 1975; Warren and Brandeis 1890). Over the years, a range of technological advances in surveillance technology have emerged. Many of them are now integrated into daily life. Without fail, each development prompts a serious discussion of what privacy is and the importance of the protections that privacy affords. There are many examples of this dynamic playing out. For instance, the introduction of mass information collection via digital databases during the early years of computer

T. H. Grubesic, J. R. Nelson, *UAVs and Urban Spatial Analysis*,
https://doi.org/10.1007/978-3-030-35865-5_5

automation spawned serious privacy concerns from the general public (Nowak and Phelps 1992; Reidenberg 1991). The collection and use of user information, mostly for the purposes of marketing, brought significant public controversy and protests. In this context, people were concerned about how personal information was being used by companies for targeted advertising—a practice considered by many to be a violation of their privacy (Nissenbaum 1998). A second controversial surveillance technology, which draws many parallels to the current discussions regarding sUAS, is closed circuit television (CCTV). Widely introduced to cities in the late 1990s, CCTV was met with mixed emotions and serious privacy concerns (Fay 1998; Sousa and Madensen 2016). While some felt it added an important layer of protection and security for the general public, others believed it would be used, disproportionately, to monitor minority populations and collect personal information (Fyfe and Bannister 1996).

In the late 1960s and mid-1970s, there was a burst of important research on the implications of emerging technologies for privacy (Altman 1976, 1977; Westin 1966, 1968). Both Altman and Westin developed an operating definition of what privacy is (or should be), as well as a series of reflections pertaining to how the new wave of surveillance technologies were being leveraged to collect and use information about the general public. Westin (1968) defined privacy as the voluntary and temporary withdrawal of a person from general society through physical or psychological means. For Westin, privacy involved barriers between oneself and the rest of the world. Later, Altman (1975) defined privacy as the selective control of access to the self. This still involved the idea of a barrier, but also the notion of being able to control the amount of information that an individual chooses to share with others. These early definitions remain relevant today because they apply to data protection in the broadest sense (including photos or videos taken of people and their property). Collectively, these definitions describe both active and passive mechanisms that denote private space and what people choose to share about themselves with an external audience (Bennett 1992). There are, of course, formal rules and regulations in place to passively protect individual privacy. However, there are also active methods to delineate private space. Whenever a technology comes along and challenges these traditional institutions, contestation soon follows.

A simple way to think about this, at least for citizens of the United States (and many other places), is that privacy is a civil liberty that grants citizens the right to be left alone and guarantees freedom from unreasonable searches of house and person (Amar 1994). Yet, the Fourth Amendment, common law, and statutory law are limited by their language, and an exhaustive federal privacy law does *not* exist (Knapp and Soylu 2013). With the emergence of the digital age and an increasing drive to collect and use personal information by corporations and the government, the maintenance and guarantee of the right to privacy is complicated. Consider, for example, the handling of user data by Facebook. In addition to being complicit in the Cambridge Analytica scandal (Granville 2018), Facebook routinely acquiesces to government and law enforcement requests for distributing personal data. Fighting terrorism is one of the usual excuses (Price 2018). In the United States, the only limits/protections from the distribution of personal information on the Facebook platform are unopened inbox messages that are less than 181 days old (Price 2018). To access this material, governments need a warrant and probable cause.

But Facebook is not alone. The hacking of Yahoo, Experian, and Equifax revealed the personally identifiable information of hundreds of millions of people (Timberg et al. 2017). In a nutshell, the problem is that all of these companies collect and store massive amounts of sensitive digital information, specific to individuals. This includes social security numbers, purchasing habits, credit ratings, loan information and many other forms of transactional data. For the most part, people (knowingly or unknowingly) allow this unfettered collection of personal information to occur. However, when there is a breach, or a questionable use of these data, people are forced to reconsider the meaning of privacy, as well as who should be given access to their personal information.

Many people consider privacy a social good (Johnson 2007; Kasper 2007), and there are diverse strategies for obtaining it. For example, some individuals move to gated communities, hire security, install cameras, or propose new homeowner's association mandates to make their community more exclusive. Conversely, communities failing to leverage such tactics may be subject to higher rates of private surveillance (Google Street View) and public surveillance by law enforcement (Traub 2000). This brings up certain ethical questions regarding the right to privacy. For example, when CCTV was emerging, justice scholars pointed to the potential for unequal rates of surveillance, a sentiment echoed by the public. Specifically, a survey on the ethical use of CCTV found three quarters of respondents agreeing that it would be used to monitor neighborhoods with a diverse ethnic makeup at a higher rate than their counterparts (Fyfe and Bannister 1996).

CCTV was also touted as the next great crime fighting tool. This pitch softened concerns for the public, especially among white communities (Bennett and Gelsthorpe 1996). Is it any wonder that similar language is being used to describe UASs (Engberts and Gillissen 2016)? For CCTV, studies reported small but significant numbers of individuals who were concerned that controllers of CCTV would justify the cost and demonstrate the worth of CCTV by over-scrutinizing certain groups such as black males and "scruffy people" (Bennett and Gelsthorpe 1996). Several other studies found evidence to suggest that some CCTV controllers were disproportionately suspicious of minority groups who were targeted at a higher rate than their majority counterparts (Bulos and Sarno 1996; Norris and Armstrong 1997). One could speculate that part of this was due to the higher crime stigma associated with low-income and ethnically diverse areas. But this is not universally true. Making these types of assumptions only further marginalizes communities who may already feel uneasy with these advances.

In many scenarios, sUAS could be considered a mobile CCTV cameras. Many systems have some sort of camera and a closed video feed transmitted back to the operator. As was the case with actual CCTV, the growth and use of sUAS bring about similar ethical and constitutional considerations regarding the right to privacy (West and Bowman 2016). Specifically, there are important questions pertaining to who is operating the sUAS, why it is flying, and where it is collecting data—all of which merit a deeper discussion regarding the integration of sUAS into society. To date, a large portion of the current social science research on sUAS is on the legality of their use, Fourth Amendment rights, sUAS regulation, and assurances of privacy

(Marshall 2016; Matiteyahu 2014). Other work has focused on developing an understanding of privacy from the perspective of individuals, many of whom may or may not be familiar with sUAS technology (Finn and Wright 2016; Nelson et al. 2019; Wang et al. 2016). At the core of this work is an emphasis on the factors triggering privacy concerns and their relevance.

5.2 Privacy and sUAS

At the moment, there are three distinct areas of sUAS privacy research for urban settings: (1) commercial/professional operation, (2) government operation (law enforcement, in particular), and (3) recreational users. There is some debate regarding which group poses the greatest risk to privacy, but recent work in this domain suggests that trust matters (Finn and Wright 2016; Herron et al. 2014; Nelson et al. 2019). Specifically, operator profiles matter. For example, a law enforcement sUAS will not insight the same privacy concerns as a recreational sUAS, even if privacy remains central to the discussion. Instead, it simply reflects a different facet of the concern. This is why a substantial amount of work—as well as regulation—is specific to the type of sUAS operator.

5.2.1 Commercial Use

In a survey concerning privacy and sUAS, Finn and Wright (2016) concluded that people perceived commercial sUAS operators as a medium-level threat to privacy, data protection, and ethics. Nelson et al. (2019) found similar results when asking respondents about their level of agreement with the following statement: "I am concerned about my privacy being violated by ∗user group∗." Interestingly, fewer respondents agreed with this statement regarding commercial users than they did with government sUAS operators (see Sect. 5.2.2). Furthermore, when detailing with the potential social impacts of commercial sUAS use, Rao et al. (2016) suggest that the potential for stealing sUAS data once it is collected (i.e., by hackers) would be the largest threat to individual privacy as opposed to the act of collecting the data itself.

In sum, these results suggest that the public views commercial sUAS operators as a modest risk, but there are nuances that should be considered. As detailed earlier in this book, commercial operators are required to take and pass knowledge tests, as well as follow FAA rules. Not doing so risks their license and livelihood. As a result, there is an assumption that these rules will be followed. Whether true or not, it appears that commercial operators maintain some level of trust among the public. Still, a bibliometric analysis of the sUAS literature reveals several areas of commercial privacy-related concerns, namely that commercial sUAS use can disrupt our notion of what privacy means (as CCTV and digital data collection did), that the meaning of privacy changes from culture to culture, and not enough is known (at the time of the review) regarding the relationship between privacy and the integration of commercial sUAS into everyday life (Luppicini and So 2016).

In addition, there are significant concerns regarding sUAS operations and safety, especially as corporate users continue to advance the possibility of automated sUAS delivery systems (Divis 2015; Mogg 2016). For now, the risk of mid-air collisions remains one of the major reasons underpinning the FAA's decision to restrict the widespread rollout of package delivery operations. However, there are many additional operational and regulatory hurdles to consider for maintaining a safe airspace with sUAS, helicopters, and planes operating simultaneously. These will be discussed later.

5.2.2 *Government Use*

As detailed previously, much of the privacy literature pertaining to sUAS is conducted through the lens of Fourth Amendment rights (Matiteyahu 2014). Fourth Amendment rights describe individual freedom from government intrusion via warrantless searches (Amar 1994). In this context, the ability of sUAS to monitor and search private property from above, unbeknownst to those being surveyed below, could be problematic. In fact, the Fourth Amendment stipulates that an unreasonable search is that which cannot be done with the naked eye. Is an sUAS flying at 400 ft in public airspace while capturing images of private property *reasonable*? What if the collected imagery is used to prosecute a crime? If the sUAS is operated in public airspace and identifies illegal activity, does this qualify as a warrantless search? Taking it one step further, what would stop the government from constantly surveilling citizens with sUAS technology?

These types of complex questions are not easy to answer. More importantly, we are not legal scholars, nor is this book focused on the nuances of sUAS law. That said, fears associated with government surveillance are valid. Generally speaking, the data collection practices of the U.S. government are absolutely atrocious (Greenwald 2013), and it would not be surprising if public sentiment regarding these practices transferred to sUAS technology. This, in fact, is exactly what Finn and Wright (2016) found in their recent work. The public does have significant concerns about government applications of sUAS technology. The vagaries of the Fourth Amendment are not helpful in this context, especially given the track record of US courts with respect to aerial surveillance and privacy (Farber 2014). For example, the courts ruled that there is no reasonable expectation of privacy in an area that can be seen from the sky (Cavoukian 2012; Farber 2014). This includes backyards, porches, and other areas that are hidden from view when at the street level. That said, there is a good deal of uncertainty when defining privacy in this way (Volovelsky 2014). Further, acts that might be considered a violation of privacy by some, may not be a concern for others. For example, people living in a house with a fenceless backyard or no backyard at all (e.g., an apartment) may not feel that an sUAS collecting imagery above their residence violates their privacy in any way. At the same time, a family living in a home that includes a privacy fence and large shrubs may feel violated.

To provide some regulatory guidance on these types of privacy issues, there are a variety of legislative efforts occurring at both the state and the federal levels. For example, in February 2018, the New Jersey State Legislature introduced a bill that restricts the use of sUAS for surveillance purposes by law enforcement (NJ 2018). New York Bill A00280 adds similar language, restricting all sUAS use by law enforcement unless a warrant has been granted prior to any surveillance activity (NY 2019). At the federal level, the Senate and House of Representatives introduced two identical bills specific to sUAS operation. Both prohibit the use of sUAS to gather "evidence or other information pertaining to criminal conduct or regulatory violations except to the extent authorized in a warrant issued under the Federal Rules of Criminal Procedure" (HR 2012).

Although there are restrictions, readers should note that sUAS are routinely (and creatively) used in law enforcement activities, nationwide. For example, the New York Police Department recently used an sUAS to track and help arrest a man during a standoff in Brooklyn (Tracy 2019). The North Little Rock, Arkansas Police Department operates the largest public sUAS fleet in the state, using them to investigate suspicious packages, search for missing people, monitor crowds at public events, and relay information about threats and obstacles to officers on the ground (Carroll and Popham 2019). In Castle Rock, Washington, the Lower Columbia SWAT team deployed an sUAS into an apartment to locate a suspect, but the suspect hit the drone and damaged it (Talak 2019). Readers should fully anticipate that sUAS applications in law enforcement will continue to grow over the next decade.

5.2.3 Recreational Users

Until the most recent FAA reauthorization Act (FAA 2018), *recreational* sUAS operators were largely exempt from the FAA rules that applied to commercial operators. The 2018 Act more explicitly addresses issues of privacy but still leaves the regulation to the state and local governments (FAA 2018). The new rules also adjust the "model aircraft" designation used (for years) by recreational operators to sUAS. This shift in designation accomplished two things. First, sUAS are now subject to (and regulated by) all FAA rules and guidelines, including all the penalties associated with flying sUAS in a reckless manner. Second, there are now pre-flight requirements that recreational sUAS operators must pass in order to fly, all of which are also enforceable by federal law (see Chap. 4).

Concerns over the use of imagery captured by recreational sUAS operators are genuine. As detailed earlier in this book, image quality and its associated level of detail captured by many sUAS are remarkable. For example, when flying at 400 ft and moving at a rate of 35 mph, the Wingtra sUAS can obtain images with a ground sampling distance of a few centimeters or less. Even the basic eBee platform, which

Fig. 5.1 eBee image quality, ~5 cm ground sampling distance

is equipped with a substantially lower-quality camera, can obtain imagery with a 3–5 cm GSD (Fig. 5.1). This provides an incredible resource for researchers studying urban, suburban, and rural phenomenon, but can also be viewed as a very good way to "case" locations for identifying valuable property for theft. Anyone can purchase a well-equipped sUAS such as a DJI Phantom and collect video from the backyards of houses for such purposes. This, understandably, is a source of concern for much of the public. In sum, the intersection of sUAS and privacy is complex and multifaceted. Readers interested in using sUAS in urban settings should be aware of these nuances and what it means for operating an sUAS above populated urban areas.

5.3 sUAS Privacy Research

If the sUAS technology is going to be used more widely in society, it is going to be increasingly important to understand how people feel about them. Public sentiment is a strong motivator for public policy. If the public does not trust drones to be used in a manner that adheres to social norms (written or unwritten), regulations are likely going to be more, rather than less restrictive. As readers may have guessed, privacy is a top public concern. As a result, researchers devote a significant amount of effort to capturing the antecedents of privacy concerns. Through a variety of methodological approaches, from large surveys of the population to more intimate in-person experiments, research is aimed at capturing the feelings about privacy and sUAS operations. By deepening our understanding of what makes people nervous about privacy and sUAS operation, sUAS manufacturers, commercial operators, and drone proponents can work toward mitigating these concerns through operational, administrative, and policy safeguards.

Within this context, a particularly salient finding concerns the level of familiarity that an individual has with sUAS in general. The public is becoming increasingly familiar with sUAS operations and that may have an effect on how individuals feel about their use with respect to privacy. In a recent survey, the Pew Research group asked individuals several questions about sUAS operation and found that 8% of the people they interviewed owned an sUAS and more than 50% had seen an sUAS operating in person (Hitlin 2017). Fifty percent is a substantial number and speaks to the growing prevalence of sUAS throughout the United States.

One of the main hypotheses driving our recently published study on privacy and sUAS (Nelson et al. 2019) was termed the "familiarity hypothesis." Kahan et al. (2008) break this down, quite simply, stating that *as the familiarity with a technology increases, support for its use will also grow*. Given this hypothesis, we expected to see that individuals who were more familiar with sUAS would find their use *less* of a privacy risk when compared to individuals with limited familiarity of sUAS technology. Indeed, survey results from Nelson et al. (2019) revealed that respondents who were more familiar with sUAS technology rated risk to privacy as lower on average (Fig. 5.2). Mean privacy was scored on a 1–5 Likert scale while level of drone familiarity was a composite score consisting of yes (1) and no (0) questions regarding their use and knowledge of drones. Higher scores on the x-axis indicate more familiarity.

Still, as sUAS become more ubiquitous and the public's familiarity with their capability grows, an opposite reaction could manifest. That is, as people gain more knowledge about the type of information sUAS can collect, many may perceive sUAS as a greater threat to privacy (Chang et al. 2017). To be clear, seeing an sUAS

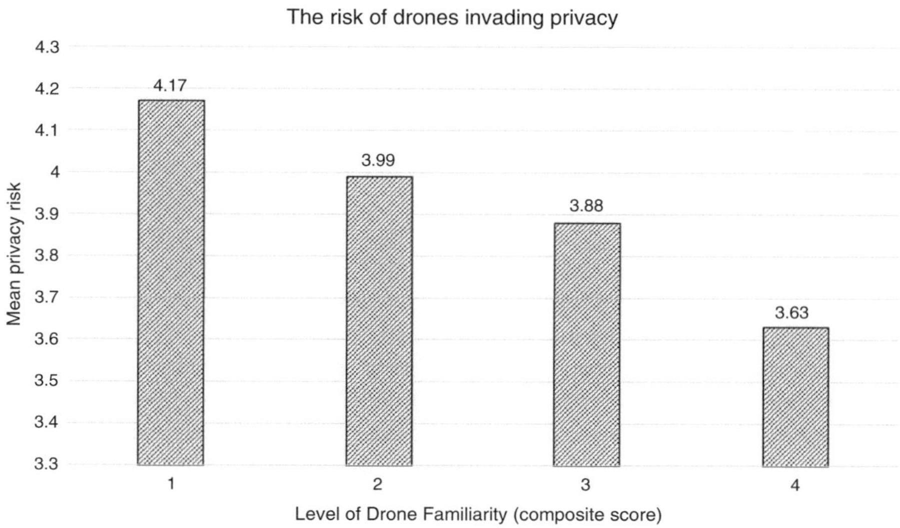

Fig. 5.2 Familiarity with drones and perceived levels of privacy risk (Nelson et al. 2019)

in action—the heights that it can fly, its size, quality of camera, and its ability to move freely about an area—can insight strong feelings regarding privacy and safety (Chang et al. 2017; Wang et al. 2016).

In addition to measuring the public's familiarity with sUAS technology, the PEW study also asked several questions about sUAS flight, including how people would feel about a drone flying near their home and where people believed a drone should be flown (Hitlin 2017). Where the former is concerned, the top three responses were curious (58%), interested (45%), and nervous (26%). For the latter, the survey findings suggested that drones should not be allowed to fly near people's homes (54%) and should only be allowed to fly in public parks (44%). These results suggest that the use of sUAS technology will likely be less contentious if the base of operations starts in a public area, such as a park, rather than somewhere more private, such as a residential area.

5.3.1 Privacy Considerations in the Field

Given the information documented by researchers regarding sUAS and privacy, there are several flight considerations worth noting. In our experience, most passersby are curious and/or interested in the operation of the sUAS—asking lots of questions about what we are doing, measuring, how much the sUAS costs, etc. This is not to say that all the people we encounter are supportive. During an effort to map local washes (channels where water flows during monsoon storms) in the Phoenix area, we established a staging area in a residential cul-de-sac and residents were uneasy. Our experiences in the field confirm what some studies have uncovered, suggesting that people will be more accepting of sUAS operations when it is done in public spaces (e.g., parks) or other locales that are considered to be part of the public domain.

Although there is a considerable literature on sUAS and privacy, the underlying factors that make the intersection of privacy and sUAS technology a provocative topic are often glossed-over. Our deep, multi-year experience in the field with sUAS in the Phoenix area and elsewhere suggests that the level of privacy that people expect is extremely variable. Most noticeable are the cultural differences in privacy expectations, as well as the social mechanisms used to regulate personal privacy. This is strongly supported by the early work of Altman (1977). For example, two of the study areas that we frequent are *very* different in social, cultural, and economic composition. One of the neighborhoods is decidedly suburban, white and middle class, while the other is more urban, poorer, and demographically more diverse. From an academic perspective, this divergence in community structure is helpful when exploring social disorder (Chap. 7) and discrepancies in access and use of solar power (Chap. 8). However, it is the differential levels of engagement that our teams experience with community members, in the field, that are the most telling. This ranges from the comments we receive to the questions that are asked of us. In the more disenfranchised neighborhood, community members seem genuinely interested about how the sUAS technology will help improve neighborhood infrastructure and quality of life, how it is programmed and its overall capabilities. In the second, more affluent neighborhood, people are warier and significantly more skeptical of its operation. We did not expect this. In fact, we anticipated the opposite reaction in both neighborhoods. On one occasion, in the more affluent locale, a passerby did not ask any questions, but simply mumbled "hellfire" as he walked by our staging area. To this day, we are not exactly sure what he meant by this, but we did nickname our eBee "Hellfire" in his honor. Are the reactions in these neighborhoods typical or generalizable? Likely not. However, it does speak to the variability that we have encountered in the field from community members. Privacy is undoubtedly important to both communities, but privacy concerns can manifest in different ways.

Consider, for example, locations in Arizona, California, Texas, or any urban area that includes a range of neighborhoods delineated by gates, security guards, homeowners associations, and privacy fences. All of these physical manifestations of privacy suggest that flying an sUAS in/around these areas will be poorly received. When scouting study locations and staging areas here in Arizona, we are increasingly cognizant of these types of privacy indicators. If our study requires that we fly near these areas, it helps when our university affiliation is clearly indicated. In fact, some of our recent works suggest that this type of organizational affiliation confers an additional level of trust between the public and sUAS operators (Nelson et al. 2019). Specifically, individuals feel that operators affiliated with a larger educational institution adhere to a formal code of ethics when conducting operations. If needed, the public believes that operators can be held accountable if something goes wrong. Again, this also supports the findings of Finn and Wright (2016) and their work pertaining to commercial operators.

With this in mind, it is important to account for the social and cultural environment in which the sUAS is operated. Just because some areas of the population cannot afford to live in a gated community does not mean they will be more welcoming to sUAS flying overhead. Researchers looking to highlight the benefits of sUAS research in low-income environments should be aware of what their presence and selection of that neighborhood looks like from the standpoint of the community members. Furthermore, tensions between communities of color and law enforcement are already high. In addition, a reiteration of social stigma by using certain areas to highlight very specific social phenomenon may incite some serious animosity (or perhaps confrontation) from the community members. If privacy is in short supply, it can be disconcerting to have an sUAS flying overhead, especially if the community is unaware of who is operating the sUAS and for what purpose (Lieberman and Miethe 2018).

5.4 Mechanisms to Mitigate Privacy Concerns

The sUAS-privacy nexus is a dynamic and changing landscape that is already playing a pivotal role in the development of sUAS technology, as well as the policies governing where and how drones can be used. There is an active field of researchers attempting to determine how to best allay these concerns through developing new operational (and technological) strategies. From an operational perspective, feedback from study participants, the community, and local government partners is valuable and must be considered when developing the parameters of research missions. There are technological strategies as well. As detailed previously, the implementation of geofences provides one way to minimize the intrusion of unwanted sUAS activity for corporate or university campuses, key infrastructure facilities and perhaps (in time) private residences. However, the logistics associated with making geofences a viable technological solution are daunting. In particular, the implementation of a geofence would require that all sUAS developers and any companies involved with mission-planning software development agree to implement privacy-protection functionality. This is no easy task. Further, even if all vendors were onboard with these types of privacy controls, there will always be a market that drives the development of countermeasures to thwart them. From an operational perspective, these types of controls, whether necessary or not, could also stymie work that relies upon sUAS operations where issues of individual privacy may be salient. For example, sUAS work that is structured to address the health of local trees, turf, or other vegetation may be limited in scope if privacy controls were widespread in urban areas.

Given the known issues associated with sUAS and privacy, there is no doubt that the public would welcome additional layers of protection. As mentioned previously, Nelson et al. (2019) suggest that operators flying under the auspices of a larger institution are often considered more responsible than a typical RPIC. Further, if there are ways to directly encode more privacy controls into the sUAS itself, the public may be more trusting of the sUAS operating within a given community. Indeed, researchers interested in privacy have proposed steps to embed privacy protection mechanisms throughout the product life cycle to ensure that every stage of development is done with the privacy of civilians in mind (Cavoukian 2012). sUAS developers and users should remain open to these types of developments.

5.4.1 *Operational Mechanisms to Enhance Privacy Assurance*

While the idea of encoding privacy controls into sUAS is both important and somewhat provocative, realizing such developments will take time. For now, sUAS users and the industry more generally rely upon policy mechanisms to police the responsible use of drone technology. For example, consider the following comment from a city council discussion regarding sUAS regulation in Hermosa Beach, California (HB 2016):

> I have had experience with drones hovering right outside our windows. We've had the privacy concerns. We don't know who it is, we haven't been able to find who it is and there's no registration number. They're hiding who knows where (and) we don't know what their intent is…the difference between the drone and somebody looking out at the beach or at the ocean with a telescope, even for whales, is that a telescope doesn't record.

This is a loaded comment that touches upon several important points regarding sUAS, privacy and mechanisms for their regulation. First, this resident is clearly uncomfortable with *not* knowing where the operator of the sUAS is located. Although the FAA does have a line of sight rule, it does not state that the operator must also be visible. Second, and relatedly, this comment touches on the ability of a passerby to be able to identify the sUAS when the operator cannot be found. In this case there was no registration number visible on the sUAS and even if there was, the stickers are small and difficult to see—especially when the sUAS is moving. Third, this comment directly acknowledges the unease of the resident with respect to the sUAS recording (e.g., video, audio, both). It seems that the resident accepts that people have an ability to "peep" with other technologies, such as a telescope, but it is the potential for an sUAS to record activity that makes its presence more problematic.

From a broader perspective, it is incumbent upon researchers to identify the aspects of sUAS activity with the potential to incite distrust, fear, and privacy violations. Simply put, the first step in understanding how to alleviate concerns is to identify both the nature and nuances of the concerns. For example, Chang et al. (2017) qualitatively recorded the reactions of study participants to an sUAS—noting how the form, sound, color, and movement impacted privacy and safety concerns. Others have provided participants with scenarios to determine when and where the use of an sUAS in an urban area *would not* be an invasion of privacy. Again, recent findings showed that commercial use is perceived as less risky, especially when the sUAS is operating in spaces considered part of the public domain (Wang et al. 2016). However, if study participants believe that the sUAS is recording personally identifiable information such as faces, car license plates and the like, respondents express more discomfort with the sUAS.

As noted earlier, efforts to address these privacy concerns include more rigorous operational guidelines and policies, as well as software and hardware modifications. Recent efforts include the development of privacy impact assessments (PIA) and sUAS product design frameworks that account for privacy at all stages of development (Cavoukian 2012). Interestingly, this research is conducted from the perspectives of both bystanders and operators because of how threats to privacy are perceived and influenced by familiarity with sUAS technology (Kahan et al. 2008). For example, Yao et al. (2017) surveyed 886 individuals; 717 were unfamiliar with sUAS operations, while 169 were experienced operators. The researchers presented the study participants with a handful of possible technical innovations that would limit the potential for personally identifiable information to be captured. They then asked respondents to rate how likely they would be to use (1) these mechanisms, (2) the perceived effectiveness of them, and (3) their practicality. Registration was rated highest for both groups and in all three response categories. The second highest was the integration and use of automatic face blurring technology. Geofences, LED light registration technology, recording indicator lights, and drone operator apps that facilitate communication were also options but not viewed as practical or as effective as the registration and face blurring.

5.4.2 Beyond Operational Controls

There are several administrative approaches to enhance privacy with sUAS that look promising, including the development of privacy impact assessments, platform designs that incorporate policy, and technical innovations to explicitly account for privacy at all stages of the product life cycle (Cavoukian 2012). The goal of this approach is to embed "privacy into the design specifications of information technologies, accountable business practices, and networked infrastructures, as the default, right from the outset" (p. 17). That is, mechanisms to ensure privacy should be pursued proactively rather than reactively, and done so from the initial development of the sUAS. Wang et al. (2016) detail mechanisms to enhance privacy in a similar manner by outlining strategies for proactive/reactive approaches for both operator and bystander. They also include high-level principles such as making the sUAS and operators more discoverable, enabling communication between operators and citizens, and making sUAS design customizable and more sensitive to local social and cultural norms.

Policy-based solutions for improving the link between sUAS and privacy are also continually under development. Legislators routinely craft new legislation that adds language to existing trespass, voyeurism, and privacy laws to account for sUAS operations (AUVSI 2018; Gruber 2015). In addition, cities are formally adopting the FAA rules for operation into local ordinances, giving their law enforcement agencies the tools required to penalize operators found to be in violation of the rules

(OC 2016). And speaking of law enforcement, there are additional policy measures delineating sUAS use for such agencies. As noted earlier, House and Senate bills introduced in 2012 explicitly restrict mass surveillance operations and represent an important first step. States are also introducing legislation to protect Fourth Amendment rights, and nongovernmental organizations, such as the American Civil Liberties Union (ACLU), have been outspoken in their support of restrictions to government use of sUAS. In fact, the ACLU has published several reports proposing guidelines pertaining to how a government should use sUAS responsibly. This includes recommendations on where, when, and how government agencies can operate sUAS, guidelines pertaining to image retention, procedures for providing public notice of sUAS use, democratic control of how they are used, and auditing and effectiveness tracking (Stanley and Crump 2011).

There are, of course, alternatives to these more collaborative efforts for ensuring a symbiotic relationship between sUAS and privacy. For example, it is possible to incapacitate sUAS, forcing them to abort missions. This includes hardware capable of disabling sUAS video feeds (Sun et al. 2017), signal jamming guns, a net wielding/deploying sUAS, and methods for remotely hacking sUAS software (Dormehl 2018). In addition, there is the relatively high-profile story of a Dutch police force that was training eagles to take-down rogue sUAS (Holligan 2016). To absolutely no one's surprise, the eagles were not thrilled about doing this job and did not behave as trained. The program was scuttled (Ong 2017).

In sum, as sUAS technology continues to grow in popularity it is likely that the policy landscape will begin to converge and a standard suite of guidelines will exist in most cities and states. Variations between states (and perhaps communities) may linger, and this should be accounted for when conducting local operations. It is also important to remember that with every FAA re/authorization act, new language is added concerning sUAS operations. The aspects of sUAS that make them so attractive to researchers and practitioners is also the reason why they are contentious, especially when it comes to privacy.

5.5 Conclusion

In this chapter we detailed some of the nuances of sUAS use and privacy, while also highlighting why privacy is important to consider prior to operating an sUAS. Privacy is one of the most cited reasons why an sUAS should not be operated in the urban environment. Privacy is one of the reasons why sUAS operations were initially banned in some cities. It is also the foundation for regulations restricting sUAS operations. This is understandable. Quite simply, people enjoy their privacy. When a technology comes along that can simultaneously maintain a foothold in public and private space, it becomes a point of contention. That said, efforts are underway to help ensure the ethical use of sUAS in the urban environment through technological innovations and operating guidelines. Irresponsible deployments of sUAS, especially those that disregard privacy, further motivates efforts to limit sUAS operation in urban areas. In fact, most incidents of sUAS razing are connected to individuals feeling that an sUAS was too close to a home and invading personal privacy (Farivar 2016; McCarthy 2015). There is no doubt that privacy is a hot-button issue for many. As sUAS deployments continue to grow, we can expect that privacy will remain an important and critical component to any regulatory or operational changes.

References

Altman, I. (1975). *The environment and social behavior: Privacy, personal space, territory, and crowding*. California: Brooks/Cole Publishing Company.

Altman, I. (1976). Privacy: A conceptual analysis. *Environment and Behavior, 8*(1), 7.

Altman, I. (1977). Privacy regulation: Culturally universal or culturally specific? *Journal of Social Issues, 33*(3), 66–84.

Amar, A. R. (1994). Fourth amendment first principles. *Harvard Law Review, 107*(4), 757–819.

AUVSI. (2018). *2018 state legislative map*. Arlington: The Association for Unmanned Aerial Vehicles International. Retrieved July 6, 2018, from http://cqrcengage.com/auvsi/statelegmap

Bennett, C. J. (1992). *Regulating privacy: Data protection and public policy in Europe and the United States*. Ithica: Cornell University Press.

Bennett, T., & Gelsthorpe, L. (1996). Public attitudes towards CCTV in public places. *Studies on Crime and Crime Prevention, 5*(1), 72–90.

Bulos, M., & Sarno, C. (1996). *Codes of practice and public closed circuit television systems*. London: Local Government Information Unit.

Carroll, S., & Popham, N. (2019). *North Little Rock police frequently use drones. What are they watching? KATV abc 7*. Retrieved April 1, 2019, from https://tinyurl.com/y4dft8r5

Cavoukian, A. (2012). *Privacy and drones: Unmanned aerial vehicles*. Ontario: Information and Privacy Commissioner of Ontario.

Chang, V., Chundury, P., & Chetty, M. (2017). Spiders in the sky: User perceptions of drones, privacy, and security. In *Proceedings of the 2017 CHI Conference on Human Factors in Computing Systems* (pp. 6765 6776). ACM.

Divis, D. A. (2015). FAA opens door to long-range drone flights, amazon prime air R&D. *Inside Umanned Systems*. Retrieved May 12, 2016, from https://tinyurl.com/k8qs8df

Dormehl, L. (2018). 7 amazing anti-drone technologies designed to swat UAVs out of the sky. *Digitial Trends*. Retrieved January 28, 2018, from https://tinyurl.com/y2g7lhu8

Engberts, B., & Gillissen, E. (2016). Policing from above: Drone use by the police. In B. Custers (Ed.), *The future of drone use: Opportunities and threats from ethical and legal perspectives* (pp. 93–113). The Hague: TMC Asser Press.

FAA. (2018). FAA Reauthorization Act of 2018. H.R.302. 115th Congress.

Farber, H. B. (2014). Eyes in the sky: Constitutional and regulatory approaches to domestic drone deployment. *Syracuse Law Review, 64*(1), 1–45.

Farivar, C. (2016). Woman shoots drone: "It hovered for a second and I blasted it to smithereens." *Arstechnica*. Retrieved from https://tinyurl.com/jyh7hpl

Fay, S. J. (1998). Tough on crime, tough on civil liberties: Some negative aspects of Britain's wholesale adoption of CCTV surveillance during the 1990s. *International Review of Law, Computers & Technology, 12*(2), 315–347.

Finn, R. L., & Wright, D. (2016). Privacy, data protection and ethics for civil drone practice: A survey of industry, regulators and civil society organisations. *Computer Law & Security Review, 32*(4), 577–586.

Fyfe, N. R., & Bannister, J. (1996). City watching: Closed circuit television surveillance in public spaces. *The Royal Geographical Society, 28*(1), 37–46.

Granville, K. (2018). *Facebook and Cambridge analytica: What you need to know as fallout widens*. New York City: New York Times. Retrieved from https://tinyurl.com/yamwa445

Greenwald, G. (2013, June). Edward Snowden: The whistleblower behind the NSA surveillance revelations. *The Guardian*. Retrieved from https://tinyurl.com/hhvjmzn

Gruber, R. H. (2015). Commercial drones and privacy: Can we trust states with 'drone federalism'? *Richmond Journal of Law & Technology, 21*(4), 14.

HB. (2016). An ordinance of the city of Hermosa Beach 16-0174 adding a new chapter 9.38 to the Hermosa Beach municipal code to regulate unmanned aircraft, drones. *Hermosa Beach*. Retrieved from https://tinyurl.com/y35l4hb8

Herron, K. G., Jenkins Smith, H. C., & Silva, C. L. (2014). *US public perspectives on privacy, security, and unmanned aircraft systems*. Norman: University of Oklahoma.

Hitlin, P. (2017, December 19). *8% of Americans say they own a drone, while more than half have seen one in operation*. Washington, D.C.: Pew Research Center.

Holligan, A. (2016). Dutch police fight drones with eagles. *BBC News*. Retrieved April, 1, 2019, from https://tinyurl.com/y59qjszs

HR. (2012). *Preserving Freedom from Unwarranted Surveillance Act of 2012*. S.3287. 112th Congress.

Johnson, D. G. (2007). *Computer ethics*. Hoboken: Wiley.

Kahan, D. M., Braman, D., Slovic, P., Gastil, J., & Cohen, G. (2008). Cultural cognition of the risks and benefits of nanotechnology. *Nature Nanotechnology, 4*, 87.

Kasper, D. V. S. (2007). Privacy as a social good. *Social Thought & Research, 28*, 165–189.

Knapp, K., & Soylu, A. (2013). Technology: The good, the bad, and the ugly. How technology is affecting employee privacy, work life balance, and workplace relationships. *Mustang Journal of Management and Marketing, 2*, 69–80.

Lieberman, J. D., & Miethe, T. D. (2018). The thin blue line meets the big blue sky: Perceptions of police legitimacy and public attitudes towards aerial drones. *Criminal Justice Studies, 31*(1), 18–37.

Luppicini, R., & So, A. (2016). A technoethical review of commercial drone use in the context of governance, ethics, and privacy. *Technology in Society, 46*, 109–119.

Marshall, D. (2016). "What a long strange trip it's been": A journey through the FAA's drone policies and regulations. *DePaul Law Review, 65*(1), 6.

Matiteyahu, T. (2014). Drone regulations and fourth amendment rights: The interaction of state drone statutes and the reasonable expectation of privacy. *Columbia Journal of Law and Social Problems, 48*, 265.

McCarthy, C. (2015). N.J. man faces prison for shooting down drone, cops say. *NJ*. Retrieved March, 26, 2018, from https://tinyurl.com/y6y9a46y

Mogg, T. (2016). Amazon's delivery drone team will like the sound of this significant FAA decision. *Digitial Trends*. Retrieved August 15, 2019, from http://www.digitaltrends.com/cool-tech/amazon-faa-drone-rules/

Nelson, J., Grubesic, T., Wallace, D., & Chamberlain, A. (2019). The view from above: A survey on the public's perception of unmanned aerial vehicles and privacy. *Journal of Urban Technology, 26*(1), 83–105.

Nissenbaum, H. (1998). Protecting privacy in an information age: The problem of privacy in public. *Law and Philosophy, 17*(5), 559–596.

NJ. (2018). *An act concerning police surveillance and supplementing title 2A of the New Jersey statutes* (p. 1). New Jersey: State Legislature.

Norris, C., & Armstrong, G. (1997). *The unforgiving eye: CCTV surveillance in public space*. Hull: University of Hull.

Nowak, G. J., & Phelps, J. E. (1992). Understanding privacy concerns: An assessment of consumers' information-related knowledge and beliefs. *Journal of Direct Marketing, 6*(4), 28–39.

NY. (2019). *Imposes limitations on the use of drones within the state*. A00280. 2019-2020 NY General Assembly.

OC. (2016). Drones: Know before you fly. *Santa Ana*. Retrieved from https://tinyurl.com/y2n9cpux

Ong, T. (2017). Dutch police will stop using drone-hunting eagles since they weren't doing what they're told. *The Verge*. Retrieved March 15, 2019, from https://tinyurl.com/yabr2qtu

Price, D. (2018). *4 reasons why facebook is a security and privacy nightmare*. Retrieved April 5, 2019, from https://www.makeuseof.com/tag/facebook-security-privacy-nightmare/

Rachels, J. (1975). Why privacy is important. *Philosophy & Public Affairs, 4*(4), 323–333.

Rao, B., Gopi, A. G., & Maione, R. (2016). The societal impact of commercial drones. *Technology in Society, 45*, 83–90.

Reidenberg, J. R. (1991). Privacy in the information economy: A fortress or frontier for individual rights. *Federal Communications Law Journal, 44*, 195.

Sousa, W. H., & Madensen, T. D. (2016). Citizen acceptance of police interventions: An example of CCTV surveillance in Las Vegas, Nevada. *Criminal Justice Studies, 29*(1), 40–56.

Stanley, J., & Crump, C. (2011). *Protecting privacy from aerial surveillance: Recommendations for government use of drone aircraft*. New York: ACLU.

Sun, A., Gong, W., Shea, R., Liu, J., Liu, X. S., & Wang, Q. (2017). Drone privacy shield: A WiFi based defense. In *28th Annual International Symposium on Personal, Indoor, and Mobile Radio Communications* (pp. 1–5). IEEE.

Talak, C. (2019). Police, SWAT standoff ends in Castle Rock. *The Daily News*. Retrieved April 1, 2019, from https://tinyurl.com/seflqew

Timberg, C., Dwoskin, E., & Fung, B. (2017). *Data of 143 million Americans exposed in hack of credit reporting agency Equifax*. Washington, D.C.: Washington Post. Retrieved from https://tinyurl.com/y6yd3vjs

Tracy, T. (2019). NYPD uses drone to keep tabs on barricaded man during standoff. *Police One*. Retrieved April 1, 2019, from https://tinyurl.com/y5atmn39

Traub, A. M. (2000). The wall is down, now we build more: The exclusionary effects of gated communities demand stricter burdens under the FHA comment. *John Marshall Law Review, 34*(1), 379–408.

Volovelsky, U. (2014). Civilian uses of unmanned aerial vehicles and the threat to the right to privacy–an Israeli case study. *Computer Law & Security Review, 30*(3), 306–320.

Wang, Y., Xia, H., Yao, Y., & Huang, Y. (2016). Flying eyes and hidden controllers: A qualitative study of people's privacy perceptions of civilian drones in the US. In *Proceedings on Privacy Enhancing Technologies* (pp. 172–190).

Warren, S. D., & Brandeis, L. D. (1890). The right to privacy. *Harvard Law Review, 4*(5), 193–220.

West, J. P., & Bowman, J. S. (2016). The domestic use of drones: An ethical analysis of surveillance issues. *Public Administration Review, 76*(4), 649–659.

Westin, A. F. (1966). Science, privacy, and freedom: Issues and proposals for the 1970's. Part I--the current impact of surveillance on privacy. *Columbia Law Review, 66*(6), 1003–1050.

Westin, A. F. (1968). Privacy and freedom. *Washington and Lee Law Review, 25*(1), 166.

Yao, Y., Xia, H., Huang, Y., & Wang, Y. (2017). Privacy mechanisms for drones: Perceptions of drone controllers and bystanders. In *Proceedings of the 2017 CHI Conference on Human Factors in Computing Systems* (pp. 6777–6788). ACM.

Chapter 6
sUAS Data Integration for Urban Spatial Analysis

Abstract Using sUAS data for urban spatial analysis poses a variety of challenges. From basic, short-term concerns such as formatting, portability, and compatibility with geographic information systems (GIS) to more complex tasks associated with incorporating ground control points (GCPs) and adding supplementary geographic base files for analysis. The purpose of this chapter is to highlight the most efficient strategies for integrating sUAS data with other sources of urban information, including census and cadastral data, as well as a variety of urban/environmental databases typically available for metropolitan locales.

Keywords Representation · Attributes · Data models · Vector · Raster · Geodesy · Projections · Ground control points · Global positioning system · Real-time kinetics · Post-processed kinematics · Data integration · Census · ZIP codes · Cadastral · Amenities

6.1 Introduction

Before we delve into the details of sUAS data integration, it is important to lay a basic foundation (i.e., vocabulary, concepts) for readers who have limited (or very little) training in the spatial sciences. Given the space constraints of this chapter and book, there are far too many topics to cover completely, but outlining key concepts concerning representation, data models and structures, geodesy, and map projections will help readers understand the important knowledge domains associated with harmonizing disparate geographic data sources for use with sUAS data. Readers wanting more details on any one of these topics should consult the embedded references. All of these references are classics, and all will help paint a more detailed picture of these important concepts.

T. H. Grubesic, J. R. Nelson, *UAVs and Urban Spatial Analysis*,
https://doi.org/10.1007/978-3-030-35865-5_6

6.2 Foundational Concepts for sUAS Data Integration

6.2.1 Geographic Representation

Geographic representation is concerned with objects, features, and interactions that occur at the earth's surface or near the surface. These features tend to be complex, which forces analysts to create abstractions. These abstractions are simplified representations of objects, features, and interactions that make data transformations, analyses, and visualizations more efficient.

As detailed by Longley et al. (2005), all geographic data are built from "atomic" elements that represent basic facts about the geographic landscape. In their most primitive form, atomic elements serve to link a place and time with some descriptive property or attribute. For example, consider the following statement: "this chapter was written on July 11th, 2019 in a house located at latitude 48.051643 and longitude –123.309347." Although simple, it ties together location, time, and an attribute. In this instance, both the location and time can vary, but so too can the attribute. For example, one could easily document the cloud cover, temperature, or something else (other than writing a book chapter) associated with that place and time. The ability to produce and abstract atomic elements for the geographic landscape is critical for geographic representation and linking sUAS data with supplementary information.

Another challenge associated with geographic representation is dealing with different types of attributes. There are five basic categories of attribute types: (1) *nominal*, (2) *ordinal*, (3) *interval*, (4) *ratio*, and (5) *cyclic*. *Nominal* data include names or identification numbers to differentiate between instances in a database. For example, instances #10223 and #10233 refer to addresses along 42nd St. in Phoenix, Arizona. These represent only two of several hundred address instances on the street. In fact, the label "42nd St." is also nominal, referring to one of Phoenix's many streets. Neighborhood designations and land use codes represent other types of nominal data for a city. *Ordinal* data represent information that has some type of natural order. For example, if the number of amenities (e.g., grocery stores, public transport stops, bicycle paths, etc.) were calculated for 30 neighborhoods in Phoenix, then each neighborhood was ranked based on the total count of amenities from highest (1) to lowest (30), the rankings are considered ordinal. *Interval* data includes information where the differences between two values is meaningful, but there is no clear meaning associated with values of zero. The classic example is temperature. The difference between 35 and 40 °C is the same as the difference between 0 and 5 °C. Again, 0 °C has no special meaning, other than representing the freezing point of water. This is not true for *ratio* data, where defining values of zero is crucial. Elevation, distance, weight, and area are all examples of ratio data. For example, we know that a value of zero for elevation corresponds to sea level. We also know that a value of zero for weight means that an object is weightless. Finally, *cyclic* data include information on flows and their compass direction (Chrisman 1997). These data require special treatment because a compass has 360°, but the value that follows 359° is 0°. This makes the calculation of mathematical averages and other metrics more difficult. Readers interested in more information on these circular and/or directional data should consult Brunsdon and Singleton (2015) and Corcoran et al. (2009).

6.2.2 Data Models

Efforts to organize trillions of features on the earth's surface and their many attributes require data models (Goodchild 1992). The two most prevalent models for geographic data are (1) *object* and (2) *field* (Couclelis 1992).[1] The object view considers features as a series of discrete objects, located somewhere in geographic space, that consist of points, lines and areas. When combined, these individual features form abstractions of real-world entities (e.g, lakes, mountain ranges). In turn, these entities help represent relationships (e.g., distance and proximity) or variations in attributes (e.g., size). The compositional elements of the object view are usually stored in a geographic information systems as coordinates (e.g., latitude and longitude) and topology, which references the way in which features are interrelated and arranged. Figure 6.1 illustrates an example set of entities for the object view.

The field view treats geographic information differently. Instead of representing features as discrete objects and entities (see the object model), the field view consists of properties that continuously vary across space. For example, elevation is a continuous geographic variable within the field view. This is true for elevations both above and below sea level. It is also possible to generate binary classifications (e.g., 0,1) for field representations. Specifically, when certain conditions are met (e.g., presence of water), cells receive a value of one. When conditions are not met (e.g., no water), cells receive a value of zero. Figure 6.2 illustrates an example of an urban neighborhood conceptualized using the field view. Specifically, Fig. 6.2a is a digital terrain model (DTM) for a portion of the Sunnyslope neighborhood in Phoenix Arizona, with Norton Park serving as the focal point. In this example, elevation is continuous, represented by a series of raster cells and varying between 391 and 405 m above sea level. Compare this to Fig. 6.2b which displays a contour map generated from the DTM. In this instance, polylines are used to represent (and connect) interpolated positions between raster cell centers to identify lines of constant elevation. In this instance, the contour intervals are 0.50 m.

One interesting aspect to Fig. 6.2 is that raster values (i.e., field data) can transform into vector contours (i.e., object data) for geographic representation. Information is lost in the process, and one introduces elements of uncertainty and error, but the underlying question remains the same: Is the geographic representa-

[1] Readers should know that this book provides a cursory overview of these models, at best. There are many nuances to both object and field data models, including elements of error and uncertainty, that are important for representing geographic information. For more details, readers should consult Unwin (1995), Goodchild et al. (2007) and Liu et al. (2008), among others.

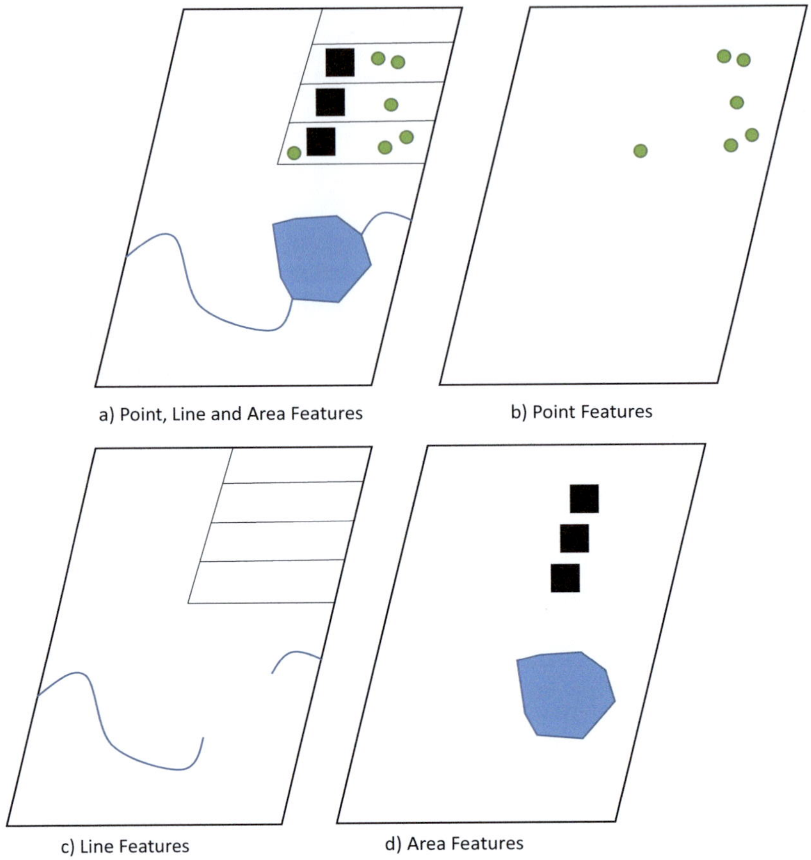

Fig. 6.1 Compositional elements of the object view

tion of elevation in Fig. 6.2 of sufficient accuracy and precision to provide a good abstraction of real-world phenomena? There is no easy answer to this question, as it will largely depend on the needs of a particular application. However, when assessing this question, it is important for readers to understand the strengths and weaknesses of raster and vector data for geographic representation.

6.2.3 Vector Data

Vector data consist of points, lines, and polygons, all of which are central to the object view. However, what readers may not realize is that the operational foundation for all lines and polygons are points. Initially, this may seem odd—after all, how can a line be a point? In truth, lines are not points, but rather a series/collection of points. Vector data uses points stored as real-world geographic coordinates to

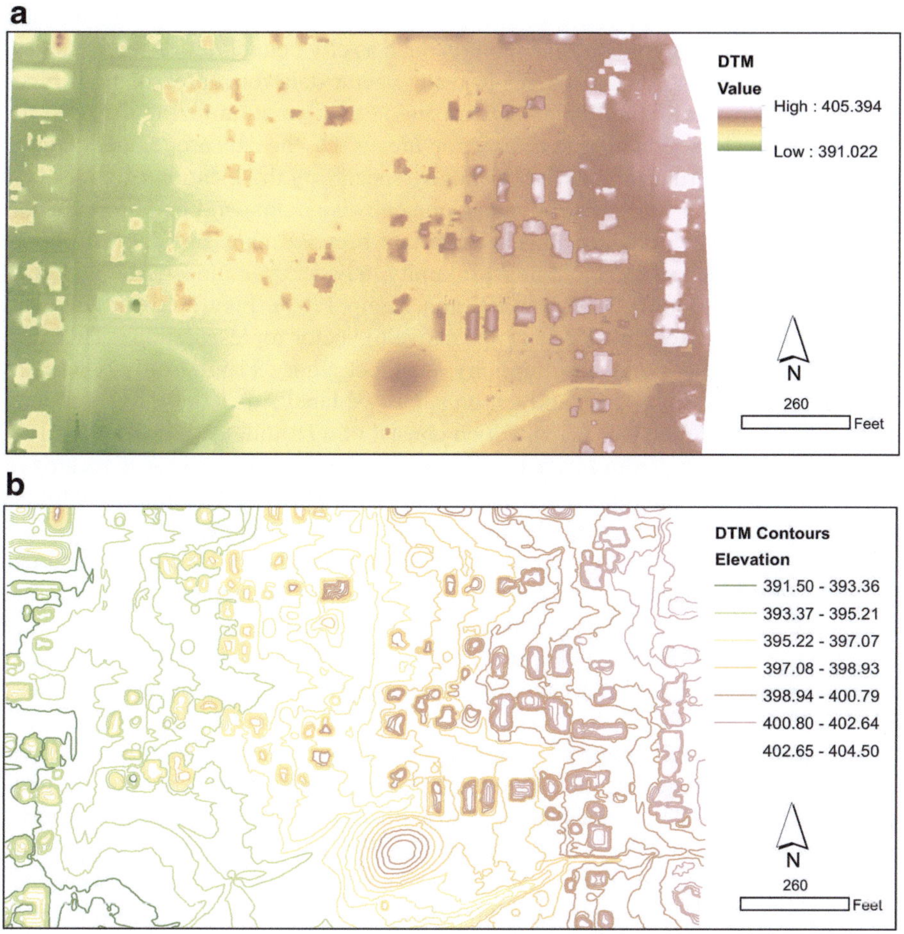

Fig. 6.2 A field view data model of elevation in the Sunnyslope neighborhood (Phoenix, AZ)

build complex objects, including lines and polygons. As detailed by Bishop and Grubesic (2016), all objects in a map that use vector data are elements of a graph. The graph (G) is defined by two sets, V and E. V is a finite set of points (i.e., vertices) and E is a set of edges (i.e., arcs or lines). G also consists of incidence relations, $V \times V$; that describe which vertices are connected by edges. When the arcs are directed, G is a directed graph. When the arcs intersect only at the vertices, G is a planar graph. The regions defined by a planar graph are *faces,* and the remaining unbounded region is an exterior face.

What should be clear to readers is that nodes are the building blocks for all vector representations. Nodes can represent point features, such as trees. Nodes can also represent the beginning and ending of all linear features in a graph, as well as the intersections of network elements (e.g., roads). The combination of nodes and arcs help build polygons, which can represent geophysical features such as swamps and mountain ranges, or administrative features such as city boundaries or service areas.

There are three major advantages to using vector data for representation. First, the data often maintain their original form and resolution, minimizing the need for generalization. This reduction in abstraction means that the data are relatively intuitive (Bishop and Grubesic 2016). Second, by specifying the location of nodes that form the vertices for lines and polygons, one can also define and build topological relationships between objects. This is important because many of the most important geographic measures, including proximity, adjacency, connectivity, and containment require topology (McDonnell and Kemp 1995). Figure 6.3 provides an example of the topological relationships native to vector representations. There are nine nodes in Fig. 6.3, but for simplicity, we label Nodes 1 and 2 only. There are three polygons (A, B, and C) and ten arcs, but we label Arc 1 only. Each node is defined by a coordinate pair (x, y), which consist of a latitude and longitude. Arc 1 is directional, starting from Node 1 and ending at Node 2. Polygon A is located to the left of Arc 1 and Polygon B is located to the right of Arc 1. This mundane topology allows one to say a variety of things about each object in the figure. For example, we know that Polygon A is adjacent to Polygon B and that Arc 1 forms a border for both. We know that Arc 1 connects to multiple arcs (not labeled). We know that Polygon C is adjacent to Polygon B. Further, if one adds a point to the middle of Polygon C, we could make a definitive statement on the containment of that point. All GIS packages that can read, interpret, and display vector data make use of these relationships.

A significant disadvantage of vector data for geographic representations is their weakness in handling continuous phenomena, such as elevation. Although the contour lines in Fig. 6.2b are reasonable approximations of the underlying elevation, readers will notice that the elevation ranges for Fig. 6.2 do not match. In part, this represents the loss of data during the conversion process from raster (DTM) to vector (contours) but also highlights how data generalization (via interpolation)

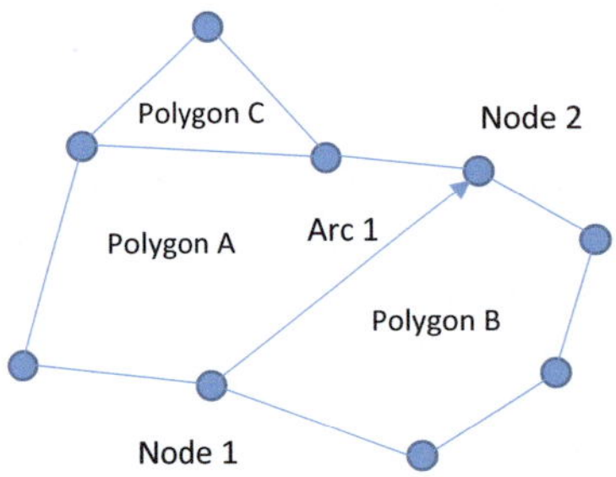

Arc	From	To	Polygon Left	Polygon Right	Node 1 *x*	Node 1 *y*	Node 2 *x*	Node 2 *y*
1	Node 1	Node 2	A	B	x	y	x	y

Fig. 6.3 Arc-node topology

increases the probability for misrepresenting elevation using vector data. Again, this might not matter for applications where errors of 1 or 2 ft are meaningless, but for high precision applications where every inch matters, this is problematic. A second disadvantage of vector data is that the drawing, manipulation, and analysis of dense vector layers can be computationally intensive. For those working with limited random-access memory (RAM) or low-grade graphics processing units (GPUs), this can limit functionality when dealing with a large number of objects in a geographic base file. Watching slower computers attempt to render large vector files is equivalent to watching paint dry.

6.2.4 Raster

Raster representations, native to the field view, do not describe discrete geographic entities or objects. Instead, rasters record and represent spatial variation by capturing the locational pattern of an attribute for a region. The spatial interval for capturing variation in these regions is typically regular, consisting of an array of cells that are square, rectangular, or some other type of geometric tessellation. These tessellations allow for the expression of geographic properties in each cell. Again, this may be something like elevation, land cover, or snowfall. Raster data include aerial photographs, satellite imagery, *drone imagery*, digital elevation models, interpolated surfaces, and many other forms of continuous phenomena.

Figure 6.4 illustrates a generic raster data structure. One major advantage of using raster representations is that very little spatial referencing information is required for grids. Unlike vector systems, where all points and polygons require explicit geographic coordinates for representation, raster grids require a single origin only. Then, based on the geographic information associated with the origin grid cell and the uniform offsets associated with other cells, locations are easily determined. Another key characteristic of raster grids is *extent*. Grid extent is a measure of the overall size of a grid. For example, in Fig. 6.4, the grid extent is 4 × 4. Although this is unitless for the moment, one could easily relabel the grid with a specific size, such as 1600 m^2. This measure of grid size relies upon the underlying resolution of the grid, which in this case is 100 m. All grid cells also have a value corresponding to a particular attribute. Values for Fig. 6.4 represent elevation. Lastly, although the organization of grids usually corresponds to rows and columns, there are alternative ways to organize and store raster data, including run length encoding, quadtrees, and/or range trees (Chang 2008). Another advantage of raster data is that each cell accommodates a single attribute only. These attributes can be nominal (e.g., trees, sand, etc.), ordinal (e.g., size of transportation routes), interval (e.g., temperature), or ratio (e.g., distance).

One of the major disadvantages in using raster representations is that intra-cell variation is lost when measuring attributes. For example, the value of 811 in Fig. 6.4 might represent the average elevation for the cell. Actual elevations within the cell likely vary. Further if one is using nominal data for a raster (e.g., wetland, forest,

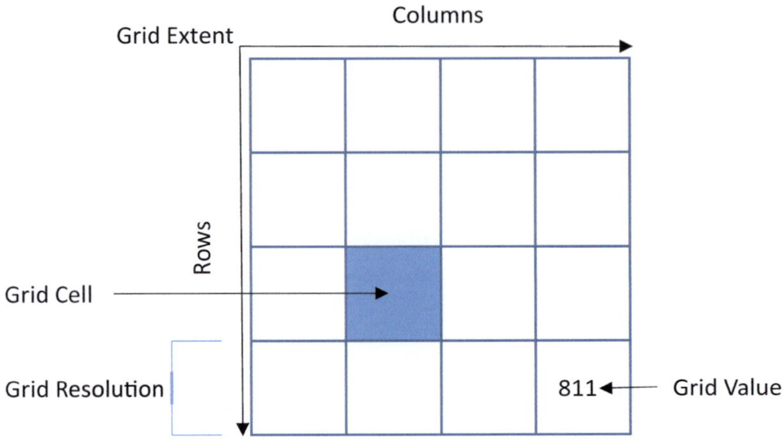

Fig. 6.4 Generic raster file structure

water), cells can suffer from the "mixed pixel problem" (Fisher 1997, p. 680). In such cases, one can apply rules for classifying cells, such as the "largest share" rule, where the dominant land cover labels a cell. Regardless of the rule chosen, application must be consistent. Another disadvantage of raster data is resolution. For most data, cell resolutions are set, and higher resolution data may not always be available. Lastly, the use of rasters to represent certain forms of geographic data, including networks or linear features, is fraught with difficulty. Depending on the resolution of the network, pixel sizes are often too large for representing the geographic nuances of stream paths, road curves, and the like. Further, the conversion between raster and vector systems for linear data can be troublesome (Bishop and Grubesic 2016).

6.3 Geodesy and Map Projections

Geodesy is the science of measuring the size and shape of the earth. Performed both globally and locally, the underlying measurements used to capture the degree to which earth is an ellipsoid can vary. For example, the satellite-derived World Geodetic System (WGS) measurements from 1972 suggested earth's equatorial radius was 6,378,135 m, while the polar radius was 6,356,750.5 m. However, the 1984 WGS suggested that the equatorial radius is 6,378,137 m, while the polar radius is 6,356,752.3 m. Although the differences between the 1972 and 1984 measurements are small, the WGS 84 ellipsoid is the global standard for mapping.

The earth is not static. Volcanos erupt, earthquakes occur, and there are variations/irregularities in earth's magnetic field. As a result, there are different geometric representations of the earth that help account for these aberrations, including spherical, ellipsoidal, and geoidal approximations. These representations form the building blocks for important functions in commercial GIS packages (e.g., distance calculations, area measurements) and image processing software. For more details on when to use, for example, the authalic sphere, versus the ellipsoid or geoid, readers should consult Robinson et al. (1995). **Hint**: the ellipsoid is the best for sUAS data.

Map projections rely upon geodetic approximations and consist of a mathematical transformation of the spherical surface of the earth into a planar surface. However, transforming something roundish (i.e., the earth) into something flat (i.e., a map) is difficult. There is no single map projection that preserves the attributes of a sphere while simultaneously maintaining shape, area, distance, and direction (Krygier and Wood 2016). At best, map projections can preserve one or two attributes, but even this is difficult. Perhaps the best way to think about the projection process is outlined by Robinson et al. (1995, p. 61). There are two stages. First, one reduces the earth's form, but it maintains its spherical shape. This "globe" exists at the same scale as the map one wants to create. Second, one transforms the three-dimensional globe into a two-dimensional surface, a map. There are additional issues associated with this transformation process that are important to consider, including scale and map distortion. Again, interested readers can consult Robinson et al. (1995) and Krygier and Wood (2016) for a full treatment of these issues.

For now, readers need to be aware of four key geometric distortions that will influence workflow, spatial analysis, and geovisualization efforts: (1) conformal, (2) equal area, (3) equidistant, and (4) azimuthal. The *conformal* property ensures the preservation of the smallest features on the earth's surface. In short, the treatment of both latitude and longitude coordinates are equal, which means that one can use conformal maps for navigation. The *equal area* property ensures that all areas drawn on a map will be proportional to those on the earth's surface. In this instance, scale adjustments occur along the meridians (north/south lines on the graticule) and parallels (east/west lines on the graticule) (Kimmerling et al. 2011). This tends to elongate, compact, and generally skew surface features, but equal area maps are excellent for calculating the size of administrative units (e.g., countries, counties). Although *equidistant* maps attempt to preserve great circle distances, this is impossible for a flat map because the actual scale varies (Krygier and Wood 2016). This means distortion for great circle distances. However, distance relationships can be maintained when using a series of lines radiating out from a single point on the map (e.g., south pole). All distance measurements along such lines represent true, great circle distances and provide a strong foundation for long-distance route planning. Finally, one can preserve direction with *azimuthal* projections. Although there is no single projection that accurately represents all directions from all locations on the earth's surface, azimuthal projections can help. They too use lines radiating from a single point (e.g., south pole) to accurately portray direction.

6.3.1 Which Projections to Use

There is no single "best" projection for urban analysis when deploying an sUAS. Rather, there are many different types of projections, each with their own specific purpose and/or strength. Here, we highlight three of the most popular (and important) projections for sUAS applications. There are many alternatives, but we cannot possibly cover all of them in this chapter. Again, readers should consult Robinson et al. (1995) for more details.

6.3.1.1 Cylindrical Equidistant (i.e., Plate Carree)

Plate Carree remains one of the most popular projections in the spatial sciences. It is equirectangular, mapping both meridians and parallels as straight lines, heavily distorting the earth's surface. However, the major advantage to Plate Carree is its ability to maintain the correct distance between portions of the map and the equator (along the meridians and parallels). In addition, it is particularly user-friendly for raster databases (e.g., NASA World Wind) because rasters fit neatly into the equirectangular grid. However, unless users reproject Plate Carree data into Cartesian coordinates, all calculations pertaining to area or direction will be inaccurate, as will many distance measurements (Bishop and Grubesic 2016). For more details on how this is problematic for spatial analysis, see Murray and Grubesic (2012).

6.3.1.2 Universal Transverse Mercator (UTM)

Universal transverse mercator (UTM) is a family of projections that include both national and global coverages. As one might gather, UTM is an equatorial projection (Mercator), but it wraps the cylinder around the poles rather than the equator. UTM is conformal, but the loxodromes are no longer straight lines.[2] Globally, there are 60 UTM zones. Within each zone, the projected north-south strops are virtually free from shape distortion and display minimal area distortion. As a result, UTM (in any flavor) is a preferred projection for conducting spatial analysis. As detailed by Longley et al. (2005), the calculation of distance between points will yield 0.04% error, or less.

6.3.1.3 State Plane Coordinate System (SPCS)

When accuracy is paramount and analysts are looking to minimize distortion, the State Plane Coordinate System (SPCS) remains the best option, especially for sUAS applications. SPCS is the standard projection for surveying applications, utility

[2]Loxodromes are imaginary lines that cross all meridians of longitude at the same angle.

placement, and cadastres. Each state in the United States has at least one unique zone but larger states can have several. For example, the state of California is so large that it has six state plane zones in the North American Datum (NAD) 1983. Projection for all of California's zones is Lambert Conformal Conic. Indiana, which is much smaller, has two state plane zones (NAD 1983) in the Transverse Mercator projection. The shape of each state drives variations in map projections. As mentioned previously, SPCS is excellent for spatial analysis and distance calculations and there are versions that use both the metric and imperial systems of measurement.

6.4 Ground Control Points (GCPs) and Global Positioning Systems (GPS)

Given the foci of this chapter thus far, readers should not be surprised that the first core topic in the sUAS data integration subsection is ground control points (GCPs). GCPs represent known locations (latitude and longitude) in the sUAS survey area. The locations display high geographic fidelity, because supplementary data on the location of GCPs are acquired either prior to or during the sUAS mission. Ground control points are important for mapping and spatial analysis because they help minimize the uncertainty and error associated with the collection of sUAS imagery. Think of GCPs as the corners of a jigsaw puzzle, helping to inform the spatial extent of a survey area, but also helping to direct where to place the other pieces within the puzzle.

It is important to note that not all sUAS projects require ground control points. If the survey team is using a high-quality drone like the eBee or Wingtra and they are comfortable with up to 1+ meter of error/uncertainty in the mapping output, GCPs are not necessary. To be clear, for many mapping projects, this level of error is not a concern. Consider, for example, that the 2010 TIGER/Line shapefiles distributed by the United States Census Bureau have a horizontal spatial accuracy of approximately 7.6 m (Census Bureau [United States] 2019). One-meter accuracy from the sUAS looks relatively good, in comparison.

Operationally, ground control points can take several forms. Many GCP options are pieces of cloth with eyelets for securing them to the ground. Consider, for example, the Sky High "Bull's-eye GCP, which is 48 × 48 in. and made of synthetic fabric. In July 2019, a six-pack of these was sold for ~$40 USD on Amazon.com. The major advantage to using this type of GCP is that they are lightweight, foldable, and easily transportable. Their durability, however, is difficult to determine. In our work, we typically use a series of homemade ground control points (Fig. 6.5). Made from a 48 × 48 in. maple-wood plank with self-adhesive kitchen tiles on one side, the total cost to construct these GCPs was about $50. Although less convenient than fabric GCPs, their weight ensures no movement. The wood/vinyl GCPs are also extremely durable, weatherproof, and inexpensive to replace should one become damaged or lost. For sUAS pilots on a budget, these are important attributes.

Fig. 6.5 Ground control point made from a maple plank and self-adhesive vinyl floor tiles

Lastly, it is important to note that expensive alternatives are available. For example, Propeller sells GCPs called "AeroPoints" that automate much of the GCP work. However, at $6000 USD for a set of ten, plus a required subscription to the Propeller Corrections Network ($600 per year, after the first year), things get expensive, quickly. The advantage to using something like AeroPoints is the automatic uploading and correction of the GCP data for use in mapping outputs (e.g., DEMs, orthomosaics). While AeroPoints may be convenient, the more traditional approach using analog GCPs is *free*.

6.4.1 Global Positioning and GCPs

Collecting GCP data using a GPS receiver is an exercise in pain. Most notably, it is extremely time consuming. If the survey requires more than about four GCPs, it is best to plan for several hours of prep work prior to flight. This includes GCP placement, collecting GPS coordinates, and recollecting the GCPs upon completion of the flight. Further, the more time GCPs spend in the field and/or survey area, the higher the probability that something interferes with the data collection process. Not only do people move them, sometimes they disappear completely (i.e., probably stolen). To this day, we are unsure why anybody is interested in these markers, but people are.

The process of collecting GPS coordinates for GCPs is straightforward. One simply places the GPS on the GCP (centered) and waits. Depending on the quality of the GPS receiver used, the ambient environmental conditions, the location of the survey, and the number of satellites available at any given time, it may take a few

Table 6.1 Structure of flat file associated with GCP geolocation information

Type	Latitude	Longitude	Elevation	Tolerance
gcp_3d	48.11629	−123.083	39.05	0.2
gcp_3d	48.11617	−123.086	36.78	0.2
gcp_3d	48.11992	−123.084	21.57	0.2

minutes to obtain a robust position. Once obtained, they should be stored digitally in a format similar to that shown in Table 6.1.

Depending on the type of software used for processing the GCP coordinates, the actual storage format may vary. However, it is absolutely necessary to include both the horizontal (lat/lon) and vertical (elevation) data for the GCP. This ensures that the GCP location is of high fidelity. Readers will notice that the coordinates in Table 6.1 are in the WGS 84 system—the standard output for most GPS systems. One can easily convert the WGS 84 data to SPCS or any other coordinate system in a GIS package.

It is important to note that there are many GPS receivers that collect coordinate data. On the hyper-expensive side, one popular setup includes a Trimble Zephyr 3 with a Geo7x handheld unit. With centimeter-level capabilities, this is an excellent kit for applications that require extreme accuracy. In July 2019, this hardware configuration would cost around $12,000–$15,000 dollars. Our team uses a less expensive configuration that includes the Bad Elf GNSS Surveyor and no external antenna. At $600, the Bad Elf not only provides 1 m accuracy, it is compatible with many post-processed kinematic solutions (more on that later). For most of our applications, we spend close to 30 min per GCP, letting the GPS "settle." This is overkill, but we want to be sure that the coordinates are as good as possible.

Increasingly, high-quality sUAS are able to use several GPS correction technologies to improve survey and mapping accuracy. This includes real-time kinematics (RTK) and post-processed kinematics (PPK). Ultimately, both RTK and PPK help accomplish the same thing as GCPs, but they do it in a more fluid, efficient and user-friendly way. By definition, kinematics refers to a branch of dynamics that deals with aspects of motion apart from considerations of mass and force (MW 2019). Where sUAS are concerned, this is important. During a flight, sUAS are in constant motion. As a result, understanding the nuances of this motion, relative to a single point on the ground, is critical for reconstructing where each image is captured.

6.4.2 Real-Time Kinematics (RTK)

Real-time kinematics provides real-time corrections to location data when the drone is flying. RTK relies on the use of a base station. The base station is a GPS receiver positioned on a known (i.e., already established latitude and longitude coordinate pair) location, that is actively collecting real-time location information from satellites. In evaluating locational accuracy, this is notable for one important reason. If the base station is located at a known point, but it is collecting data and measuring a position that is slightly different from its actual location (errors fueled by the upper atmosphere), this is a *known error*. Known errors are easily resolved. Moreover, if the base station is able to communicate with the sUAS about positioning and known errors, real-time corrections are possible, drastically improving the georeferencing information for each image.

The results of comparative work pertaining to RTK suggest high levels of geographic consistency and repeatability across multiple flights (Forlani et al. 2018; Sensefly 2017). The problem with RTK, however, is its reliance on wireless communications for GPS corrections. If the signal between the base station (or any other piece of equipment) is lost, the corrected GPS data are no longer available. This may be temporary, but bits and pieces of the corrected data are still lost.

6.4.3 *Post-processed Kinematics (PPK)*

Post-processed kinematics is a GPS correction technology applied *after* the flight. Similar to RTK, PPK relies on a GPS base station, in a known location, that is actively collecting location information from satellites. Simultaneously, the sUAS is flying its mission, collecting GPS tagged images over the survey area (Fig. 6.6). PPK corrections occur on a computer once the mission is complete.

Specifically, one corrects images from the sUAS by using the receiver-independent exchange format (RINEX) files generated by the GPS base station. RINEX files are a data interchange format for raw GPS data. When combined with the location data from the base station, known errors are accounted for in the geotagged images and the overall accuracy of the image locations are improved. The major advantage to using PPK versus RTK is that no information is lost between the sUAS and the base station during flight. Again, they are not communicating directly at any time. Further, there is no drop in quality between RTK and PPK corrections. We strongly recommend PPK workflows for urban analysis. There are simply too many obstacles (e.g., houses, buildings) in urban environments for RTK to work flawlessly, at all times. Information will be lost. Consider the recent study conducted by Pix4D (2017). A series of flights using a senseFly eBee Plus (RTK ready), with a GSD of 2.5 cm and 80% image overlap, were conducted in an urban area. The RTK made corrections for 71% of the data, with the remaining 29% classified as "RTK float." Ultimately, while the RTK processing yielded good results (8.1 cm of error), the PPK corrections were better (6.7 cm of error).

6.5 sUAS Data Integration

Many different types of data serve to compliment sUAS imagery. One's choices are largely dependent on project needs. For urban analysis, there are several classic databases that require consideration for many projects. In the following subsections,

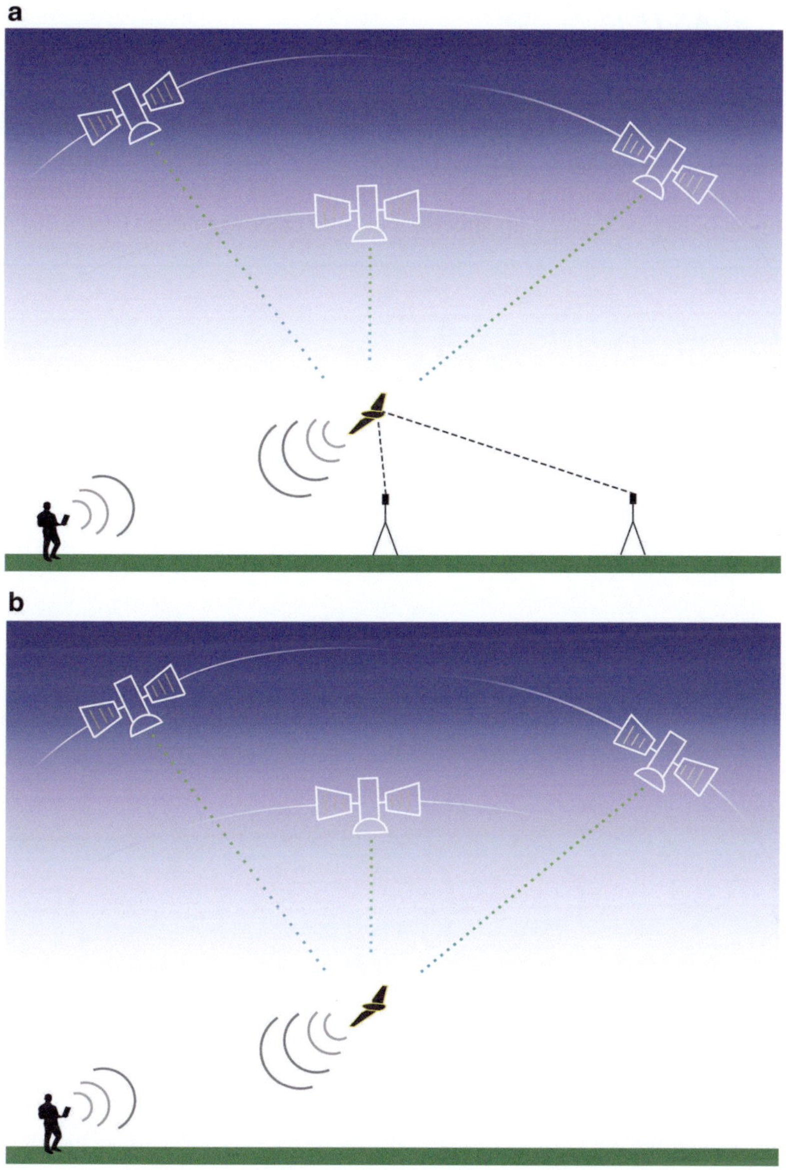

Fig. 6.6 (**a**) Real-time and post-processed kinematics for sUAS flights rely on base stations, (**b**) without base stations, the positon of the sUAS is determined by satellite links

we provide a brief overview of these data, detailing their strengths, weaknesses, and the challenges of integrating these data with sUAS imagery. While we cannot hope to cover every database for every application, readers with a background in the spatial, environmental, and socioeconomic planning sciences should be able to make the appropriate connections to their favorite supplementary data with ease.

6.5.1 sUAS Data

The remainder of this chapter uses sUAS data collected from Phoenix, AZ (March 2019) for illustrative purposes. Specifically, we use data from the Sunnyslope neighborhood in Phoenix (Fig. 6.7), which is one of the most diverse demographic and socioeconomic areas of the city. It is a neighborhood in transition, attempting to attract new businesses, clean up streetscapes, and cater to tourists (Estes 2018).

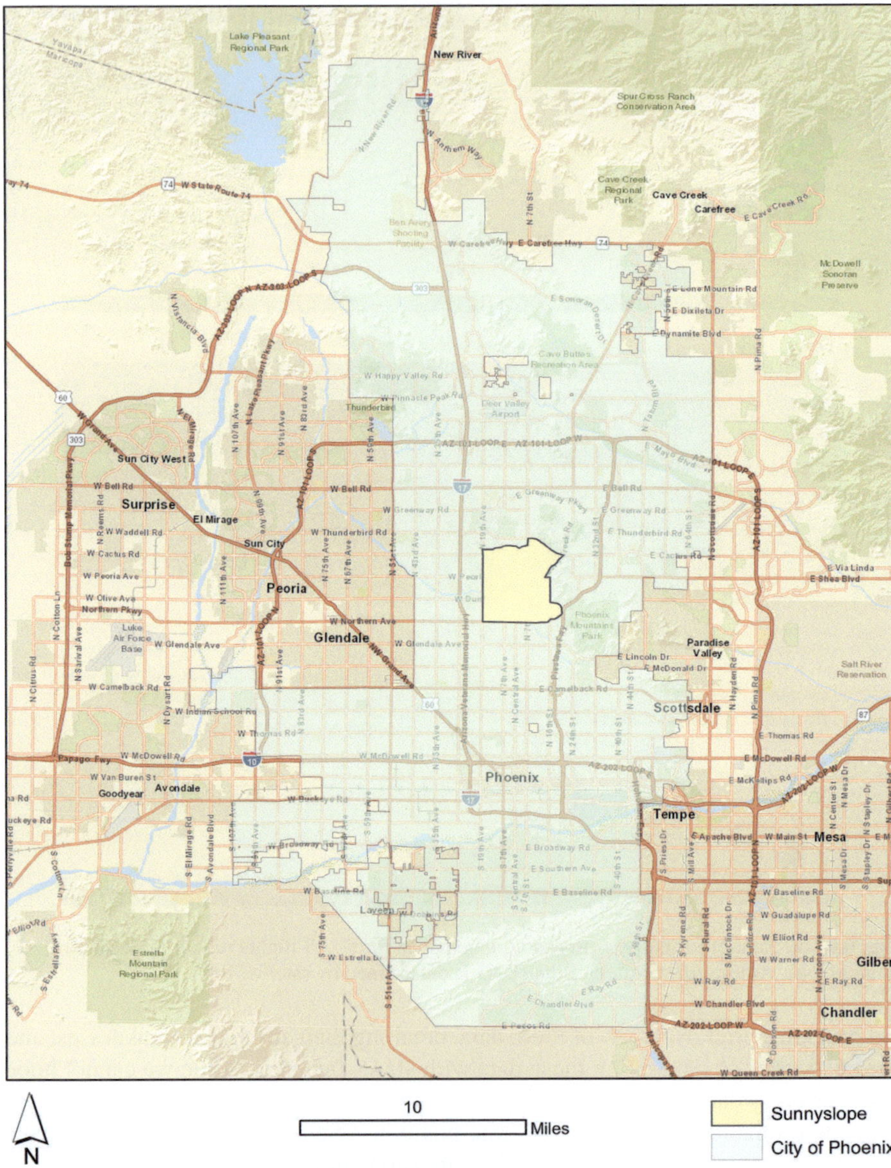

Fig. 6.7 Sunnyslope neighborhood, Phoenix, Arizona

The WingtraOne VTOL sUAS surveyed 96.53 acres (0.390 km^2), collecting 763 images with 70% overlap while flying at approximately 300 ft above ground level. The average ground sampling distance was 1.30 cm. We processed the imagery data using the Pix4D (v. 4.3.33) photogrammetry suite (Pix4D 2019) on a desktop computer running Windows 10, with an Intel Xeon E-2176M CPU at 2.70 GHz, 64GB of RAM, an Intel UHG Graphics P630 driver, and an NVIDIA Quadro P4200 GPU. The computational effort for initial processing was 35 min, plus an additional 42 min for generating the digital surface model and about 2.5 h for generating the orthomosaic.

6.5.2 Census Cartographic Boundary Files

Most readers are likely familiar with the United States Census Bureau, its decadal survey and American Community Survey (ACS) data. Both include gigabytes of information on race, education, income, and many other important data points concerning US communities and their residents. However, the operational core for most (if not all) of these data are cartographic boundary files. These files add geographical context to all of the data collected. This includes routine boundaries such as states, counties census tracts, and block groups, to more esoteric boundaries such as congressional districts, urban areas, and ZIP code tabulation areas (ZCTAs). Many, if not most, of these boundaries are far too coarse (i.e., small scale) for integrating with sUAS imagery. Most sUAS surveys take place at much larger scales. However, several census boundary files can be important for sUAS analyses, including blocks and block groups, and occasionally ZCTAs.

Census blocks represent the smallest geographic area of data collection by the Census Bureau. Boundaries correspond to streets, roads, railroads, and physiographic features (e.g., water). During 2010, there were 11,155,486 blocks in the United States and Puerto Rico. In urban areas, blocks display a relatively small geographic footprint, but in rural areas, blocks can be large and irregularly shaped. Blocks contain no socioeconomic information, but they can provide data on overall population, group quarters, housing units, and the populations of different groups (e.g., white, Asian). Figure 6.8 illustrates (in pink) the 2010 tabulation block boundaries for the Sunnyslope neighborhood in Phoenix. Figure 6.8 also illustrates (in red) the 2010 block group boundaries—which contain a variety of demographic and socioeconomic data for Sunnyslope.

The data underlying both blocks and block groups can help map spatial variations in race, population, and the like for a survey area. In turn, one can use these data, combined with urban crime data, as inputs to help explain crime rates in a neighborhood. Eventually, one can also use sUAS data to identify features that help generate crime, including alleyways, poor streetlight coverage, and other measures of physical disorder. Chapter 7 will provide a more detailed exploration of remotely sensing physical disorder for Sunnyslope.

Fig. 6.8 Census block group and block administrative areas

We hesitate to include ZIP Code Tabulation Areas (ZCTAs) as a viable geographic representation and/or unit for urban analysis, but they are such popular repositories of important data, we must. That said, readers should proceed with caution when blending sUAS data with ZCTA data. ZCTAs represent a lousy approximation of ZIP code delivery routes developed and maintained by the United States Postal Service (USPS) (Grubesic 2008; Grubesic and Matisziw 2006). At their core, five-digit zip codes do not correspond to discretely bounded geographic areas (i.e., polygons). ZIP codes are *linear features* that assign roads, streets, and addresses to postal delivery routes. Areas without the need for mail delivery do not have ZIP codes. In fact, ZCTAs represent a Frankenstein-like creation by the Census Bureau that use tabulation blocks to represent ZIP code data in the Census master address file. The process used by the Census to create the ZCTAs is tragic, at best, and full of geographic uncertainty and error. For more details, see United States Census Bureau (2019).

Figure 6.9 displays the ZCTAs for the Sunnyslope area of Phoenix. Notice that the ZCTA completely encloses the survey area. However, there may be instances where the sUAS survey area overlaps ZCTA or ZIP code boundaries, and these delineations become more important. The same goes for other units, including urban areas, congressional districts, and the like. In such instances, the ability to integrate these Census boundary data and the underlying variables of interest, in the same projection space and coordinate system, are critical.

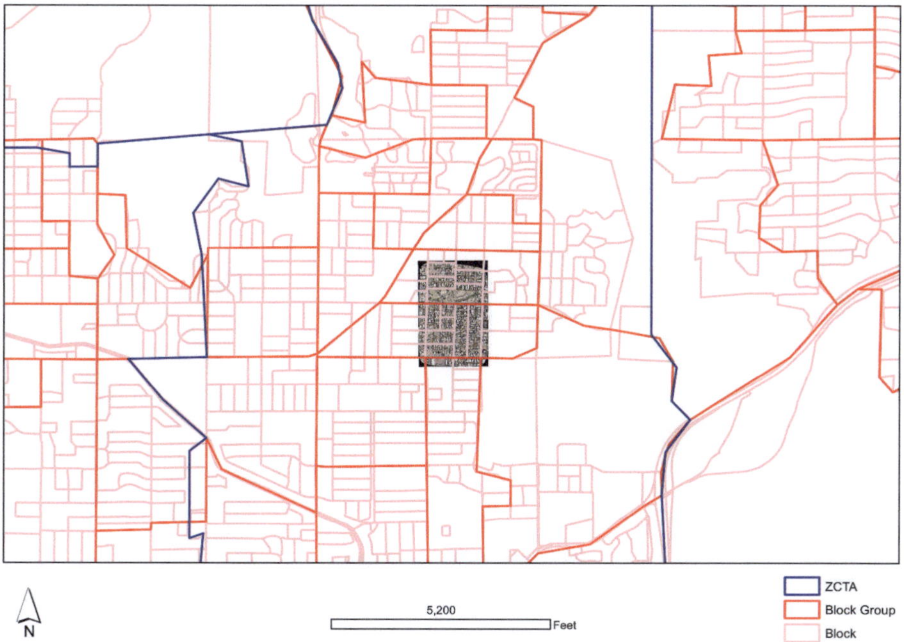

Fig. 6.9 ZCTA overlay on census blocks and block groups

6.5.3 Cadastral and Urban Structure Data

As GIS and remote sensing technologies continue to improve, the scale of urban analysis becomes more refined. Increasingly, high-resolution urban analysis relies upon both cadastral and urban structure data. "Cadastral data defines the geographic extent of the past, current and future rights and interests in real property and the spatial information needed to describe that geographic extent" (Trammell 2019). In many (if not most) locations, cadastral data also serves as the foundation for property tax assessments, as well as emergency response and 911 services. In many instances, cadastral data includes urban structure data. These building footprints provide additional information on the morphology of a community. Figure 6.10 displays two geographic base files: one for Sunnyslope parcels and the other structures.

Recent work leverages these types of data to estimate the use of water for private swimming pools (Llausàs et al. 2018), estimating urban heat islands (Li et al. 2016), estimating building energy loss (Martín-Consuegra et al. 2018), and developing more efficient and realistic emergency response routes in coastal communities (Helderop and Grubesic 2019). When cadastral and structure data are combined with sUAS imagery, new opportunities for urban analysis emerge. For example, sUAS imagery can help rapidly update urban structure data. Figure 6.11 highlights in red several buildings missing from the structure database. The algorithm may have missed these structures, or they might reflect new builds as Sunnyslope

Fig. 6.10 Cadastral and building structure data for Sunnyslope

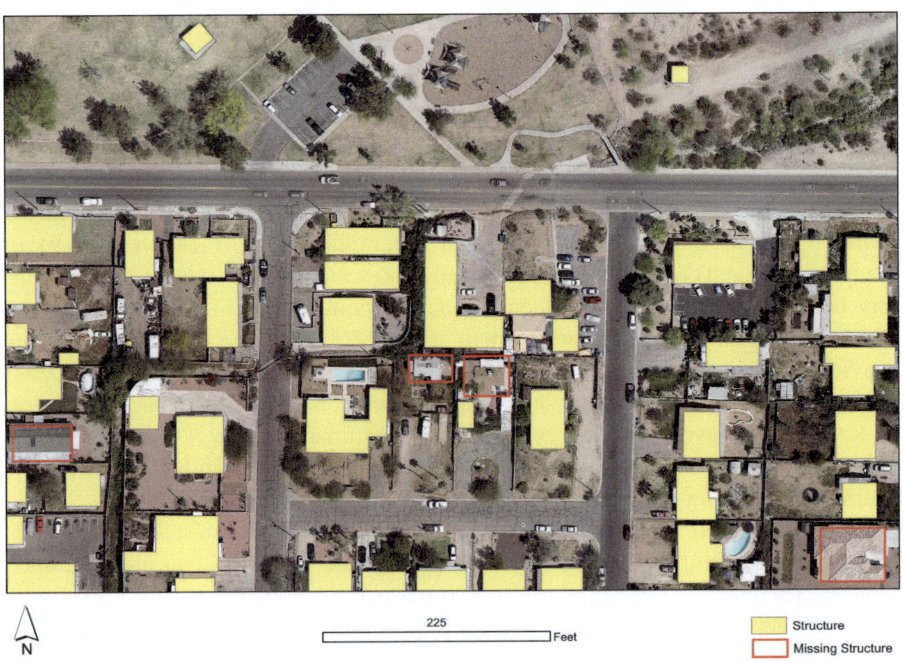

Fig. 6.11 Missing structures in the Phoenix building database (Sunnyslope) identified with sUAS imagery

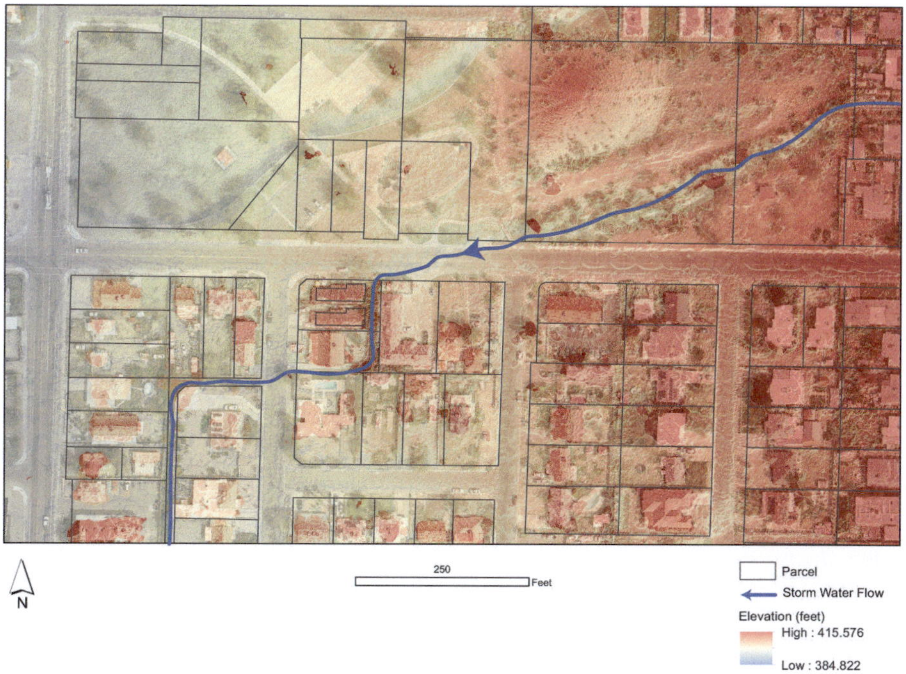

Fig. 6.12 Stormwater flows through the Norton Park wash in Sunnyslope

continues to gentrify. One can also use sUAS data to identify new solar panel installations for a structure or evaluate roofing damage after a monsoon storm. The ability to link these additional data points directly to parcels and structures is a powerful addition to real property analysis and urban management.

Finally, it is also possible to combine a digital surface model (DSM) and/or digital elevation models (DEMs) with cadastral data for examining potential stormwater flow paths via sUAS imagery (Djokic and Maidment 1991; Petrasova et al. 2017; Tokarczyk et al. 2015). Again, the addition of cadastral data are important because one can directly link the potential for dangerous stormwater flows directly to real property. Figure 6.12 displays a cartographically simplified version of the results obtained by using basic hydrological models available in ArcGIS 10.6 and the digital surface modeling capabilities of Pix4D. The blue line indicates the primary path of stormwater movement through the Norton Park wash and its ejection into a portion of the Sunnyslope neighborhood. Figure 6.13 displays a series of traditional images that highlight a range of ad hoc water management features constructed by the city of Phoenix and local residents. Notice in Fig. 6.13d, the cracked masonry wall (right), the partially sealed brick wall (left), and the general signs of dilapidation and disrepair of the infrastructure associated with keeping the stormwater away from residential properties. One can track all of these important urban issues with sUAS imagery, GIS databases, and spatial analytical approaches.

Fig. 6.13 (**a**) Looking northeast, toward Norton park. Water flows through the Norton Park was, across Hatcher Road, and into concrete channels, (**b**) The channel runs behind the parcel/house pictured in (b) and is ejected onto 12th place (**c**), (**d**) after crossing 12th place, the water is once again routed between homes into a makeshift channel/ditch until is passes into a settling basin (Palma Park) ~1500 ft downstream

6.5.4 Urban Amenities and Disamenities

Urban amenities include a long and varied list of facilities and features in the local environment that improve quality of life and personal well-being. Amenities include things such as parks, trees, community pools, restaurants, public transport stops, playgrounds, and the like (Wu 2006; Landry and Chakraborty 2009). Urban disamenities include hazardous waste sites (e.g., superfund sites), flood plains, abandoned homes, poor rubbish collection, and many other elements typically associated with urban decay (Chen and Jim 2010; Gaffney et al. 2005). In Phoenix, trees are one of the most important urban amenities, providing shade to local residents during the hottest months of the year, where temperatures routinely exceed 110 F (43 °C) (Middel et al. 2014, 2015; Bernstein et al. 2016). sUAS imagery is an excellent source for collecting and updating tree data and canopy coverage. Figure 6.14 highlights the distribution of trees, for Sunnyslope. We generated these data using the Sunnyslope sUAS imagery and a suite of basic classification routines in ArcGIS. Figure 6.14 also highlights an urban disamenity, a local floodway, associated with the Sunnyslope neighborhood (FEMA 2019).

Once the tree data are available in a point format within the GIS, additional analysis opportunities present themselves. For example, one could tabulate the total number of trees present for each parcel and calculate descriptive statistics for the neighborhood. Basic spatial statistical analysis is also possible. For example, one can calculate the local arrangement of trees for Sunnyslope using distance bands.

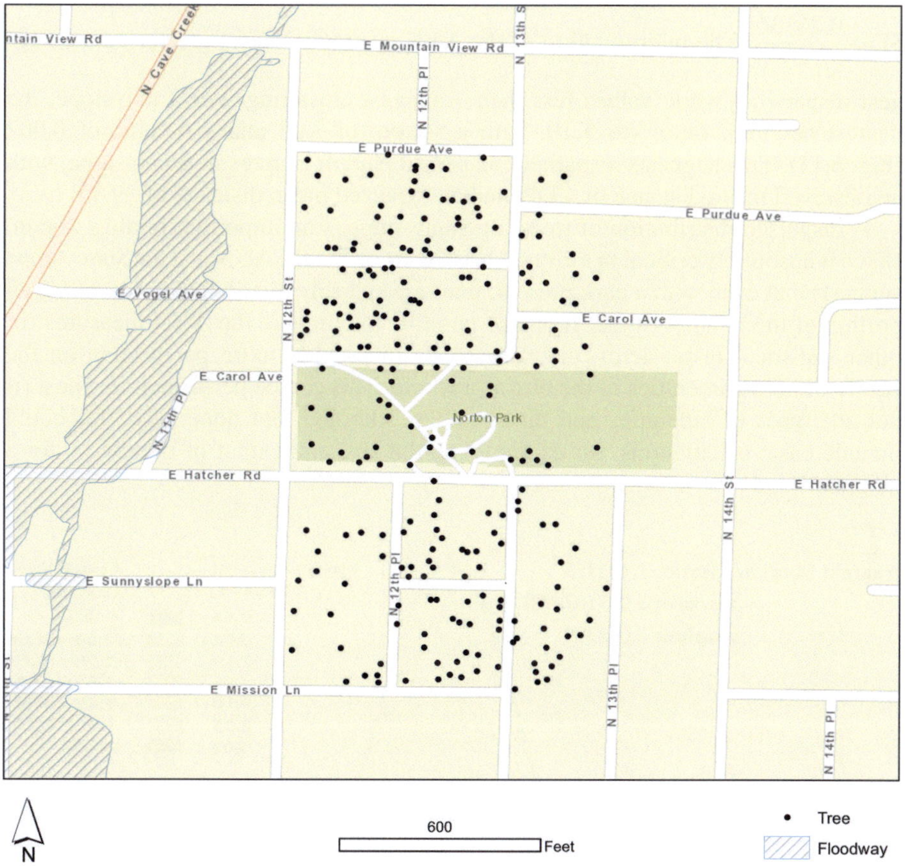

Fig. 6.14 Urban amenities (trees) and disamenities (Floodway) in Sunnyslope

In this case, the minimum distance between a tree and its ten nearest neighbors is 79.5 ft, while the average for all trees is 146.1 ft, and the maximum is 285.2 ft. A more rigorous analysis is also possible, including the calculation of average nearest neighbors. Consider the following notation:

d_i = the distance between tree i and its nearest neighbor
n = the total number of trees
A = the area of a minimum enclosing rectangle around all trees (but it can be user specified).

The average nearest neighbor (ANN) ratio is given as: $\text{ANN} = \frac{\overline{\overline{D}}\text{o}}{\overline{\overline{D}}\text{e}}$, where $\overline{\overline{D}}\text{o}$ is the observed mean distance between each tree and its nearest neighbor, $\overline{\overline{D}}\text{o} = \frac{\sum_{i=1}^{n} d_i}{n}$. $\overline{\overline{D}}\text{e}$ is the expected mean distance for the features given a randompattern, $\overline{\overline{D}}\text{e} = \frac{0.5}{\sqrt{n/A}}$. The average nearest neighbor z-score is calculated as $z = \frac{\overline{D}\text{o} - \overline{D}\text{e}}{\text{SE}}$, where

$SE = \frac{0.26136}{\sqrt{n^2 / A}}$. The interpretation of the ANN is simple.Values greater than 1 suggest dispersion, while values less than 1 suggest clustering. For Sunnyslope, the nearest neighbor ratio was 1.10, with a z-score of 3.03 and a p-value of 0.002 (Fig. 6.15). This suggests a spatially dispersed pattern of trees in Sunnyslope, with an observed mean distance of 43.49 and an expected mean distance of 39.45.

A dispersed distribution of trees for Sunnyslope is an important finding. Again, shade is absolutely critical in a city such as Phoenix, but the ANN test for Sunnyslope suggests that even with a park present, trees are sparsely distributed in the surveyed portion of the neighborhood. It is also possible to compare the ANN measures (or other statistical tests) across neighborhoods to provide more perspective on the equity of urban amenities in the city as a whole. This can be performed for most (if not all) types of amenities and disamenities. Although not done here, this could include basic calculations, for example, on the size and extent of floodways for a neighborhood.

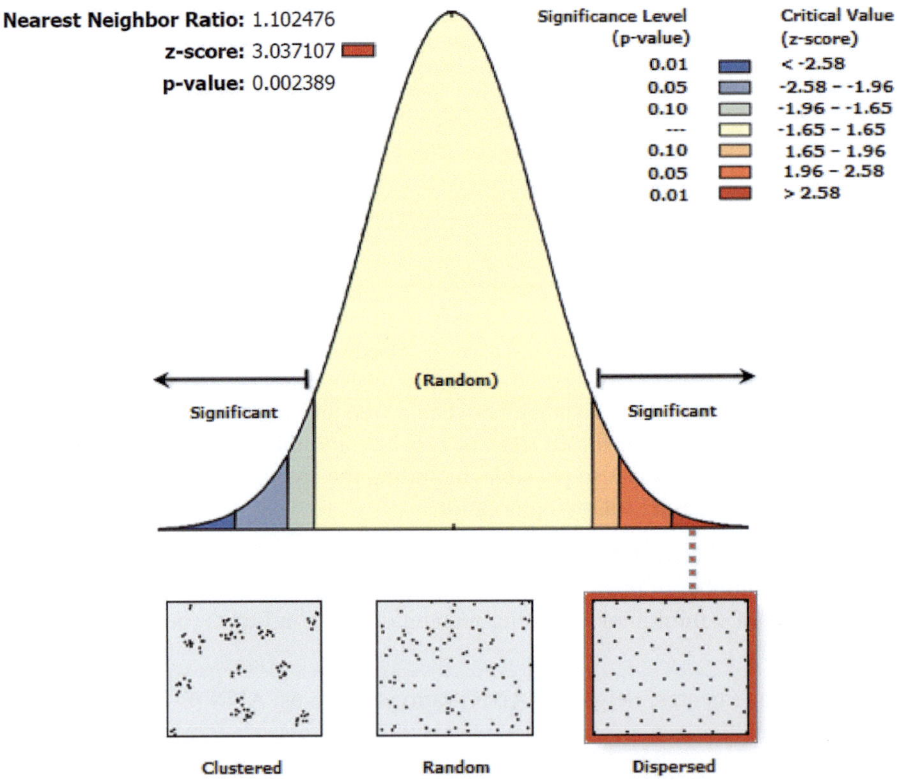

Fig. 6.15 Average nearest neighbor (ANN) test for the distribution of trees in Sunnyslope. Note: Given the z-score of 3.037, there is a less than 1% likelihood that this dispersed pattern is the result of random chance

6.6 Conclusion

We live in a data-rich world. There are far too many interesting databases to cover in this chapter, all of which one can combine with sUAS-derived imagery for urban analysis. Regardless of the data selected, analysts need to remain vigilant in how these disparate data combine for use. The purpose of this chapter was to highlight several foundational aspects of data integration, ranging from spatial representation to ground control points and global positioning systems. Readers should pay particular attention to the following takeaways. First, when dealing with spatial data representation, geographic projections, and coordinate systems matter, a lot. Harmonizing projections takes work, and the path(s) for accomplishing this task are not always obvious. Although it is tempting to believe that WGS 1984 data will automatically line up with 1983 AZ State Plane Data in feet (or whatever) in a GIS, they will not. The best way to deal with disparate projections is to establish a "working" projection at the outset of the project and to make sure to convert everything prior to analysis. This takes time, but it is critical for any work that requires an accurate spatial representation of objects (e.g., buildings), parcels, amenities, or disamenities in an urban environment.

Second, it is important that analysts pre-determine how much accuracy and precision is necessary for a project. If a few meters of error are acceptable, the use of ground control points and post-processed kinematics may not be necessary. However, when sub-meter accuracy matters, the ability to integrate GCPs and PPK becomes a priority. There are many strategies for accomplishing this and plenty of good step-by-step instructions found on the Internet (e.g., Youtube). With the right equipment, most readers should not have any problems improving survey accuracy.

Third, know the important details of all supplementary data and the way(s) in which they were collected. The brief discussion of ZCTAs covered in this chapter should serve as a wake-up call for many readers. Do not automatically assume that data are high-quality when provided by an authoritative source such as the Census Bureau. This could not be farther from the truth. All data are subject to uncertainty and error (Murray et al. 2014), and it is your job to understand exactly how that might manifest for a given project.

Finally, sUAS data will increasingly set new standards for accuracy and precision of remotely sensed urban imagery. It is crucial that analysts take care to include as much metadata (Bishop and Grubesic 2016) about surveys as possible. This goes beyond flight times and area covered. Analysts should endeavor to include information on the platforms and sensors used, how the data were processed (computers and software), and if the data are available in a public repository. Not only does this help with scientific replication, the data can also serve as a historical benchmark for urban areas that, sometime in the future, will help all of us better understand urban dynamics and the evolution of cities.

References

Bernstein, M. J., Wiek, A., Brundiers, K., Pearson, K., Minowitz, A., Kay, B., & Golub, A. (2016). Mitigating urban sprawl effects: A collaborative tree and shade intervention in Phoenix, Arizona, USA. *Local Environment, 21*(4), 414–431.

Bishop, W., & Grubesic, T. H. (2016). *Geographic Information*. Springer International.

Brunsdon, C., & Singleton, A. (Eds.). (2015). *Geocomputation: A practical primer*. Thousand Oaks: Sage.

Census Bureau [United States]. (2019). *About the 2010 census tiger/line shapefiles*. Retrieved March 16, 2019, from https://tinyurl.com/y3rgot9d

Chang, K. T. (2008). *Introduction to geographic information systems (Vol. 4)*. Boston: McGraw-Hill.

Chen, W. Y., & JIM, C. Y. (2010). Amenities and disamenities: A hedonic analysis of the heterogeneous urban landscape in Shenzhen (China). *Geographical Journal, 176*(3), 227–240.

Chrisman, N. R. (1997). *Exploring geographic information systems*. New York: Wiley.

Corcoran, J., Chhetri, P., & Stimson, R. (2009). Using circular statistics to explore the geography of the journey to work. *Papers in Regional Science, 88*(1), 119–132.

Couclelis, H. (1992). People manipulate objects (but cultivate fields): Beyond the raster-vector debate in GIS. In *Theories and methods of spatio-temporal reasoning in geographic spac*e (pp. 65–77). Berlin: Springer.

Djokic, D., & Maidment, D. R. (1991). Terrain analysis for urban stormwater modelling. *Hydrological Processes, 5*(1), 115–124.

Estes, C. (2018). Can phoenix's sunnyslope community become a tourist destination? *KJZZ*. Retrieved January 25, 2019, from https://tinyurl.com/ybcgs5j6

FEMA [Federal Emergency Management Agency]. (2019). *FEMA flood map service center*. Retrieved from https://msc.fema.gov/portal/home

Fisher, P. (1997). The pixel: A snare and a delusion. *International Journal of Remote Sensing, 18*(3):679–685.

Forlani, G., Dall'Asta, E., Diotri, F., Cella, U. M. D., Roncella, R., & Santise, M. (2018). Quality assessment of DSMs produced from UAV flights georeferenced with on-board RTK positioning. *Remote Sensing, 10*(2), 311.

Gaffney, S. H., Curriero, F. C., Strickland, P. T., Glass, G. E., Helzlsouer, K. J., & Breysse, P. N. (2005). Influence of geographic location in modeling blood pesticide levels in a community surrounding a US Environmental Protection Agency Superfund site. *Environmental Health Perspectives, 113*(12), 1712–1716.

Goodchild, M. F. (1992). Geographical data modeling. *Computers & Geosciences, 18*(4), 401–408.

Goodchild, M. F., Yuan, M., & Cova, T. J. (2007). Towards a general theory of geographic representation in GIS. *International Journal of Geographical Information Science, 21*(3), 239–260.

Grubesic, T. H. (2008). Zip codes and spatial analysis: Problems and prospects. *Socio-Economic Planning Sciences, 42*(2), 129–149.

Grubesic, T. H., & Matisziw, T. C. (2006). On the use of ZIP codes and ZIP code tabulation areas (ZCTAs) for the spatial analysis of epidemiological data. *International Journal of Health Geographics, 5*(1), 58.

Helderop, E., & Grubesic, T. H. (2019). Streets, storm surge, and the frailty of urban transport systems: A grid-based approach for identifying informal street network connections to facilitate mobility. *Transportation Research Part D: Transport and Environment, 77*, 337–351. https://doi.org/10.1016/j.trd.2018.12.024.

Kimmerling, A. J., Buckley, A. R., Muehrcke, P. C., & Muehrcke, J. O. (2011). *Map use: Reading, analysis, interpretation* (7th ed.). Redlands, CA: Esri Press Academic.

Krygier, J., & Wood, D. (2016). *Making maps: A visual guide to map design for GIS*. New York: Guilford Publications.

Landry, S. M., & Chakraborty, J. (2009). Street trees and equity: Evaluating the spatial distribution of an urban amenity. *Environment and Planning A, 41*(11), 2651–2670.

Li, X., Li, W., Middel, A., Harlan, S. L., Brazel, A. J., & Turner Ii, B. L. (2016). Remote sensing of the surface urban heat island and land architecture in Phoenix, Arizona: Combined effects of land composition and configuration and cadastral–demographic–economic factors. *Remote Sensing of Environment, 174*, 233–243.

Liu, Y., Goodchild, M. F., Guo, Q., Tian, Y., & Wu, L. (2008). Towards a General Field model and its order in GIS. *International Journal of Geographical Information Science, 22*(6), 623–643.

Llausàs, A., Hof, A., Wolf, N., Saurí, D., & Siegmund, A. (2018). Applicability of cadastral data to support the estimation of water use in private swimming pools. *Environment and Planning B: Urban Analytics and City Science, 46*(6), 1165–1181.

Longley, P. A., Goodchild, M. F., Maguire, D. J., & Rhind, D. W. (2005). *Geographic information science and systems*. Chichester, UK: Wiley.

Martín-Consuegra, F., de Frutos, F., Oteiza, I., & Agustín, H. A. (2018). Use of cadastral data to assess urban scale building energy loss. Application to a deprived quarter in Madrid. *Energy and Buildings, 171*, 50–63.

McDonnell, R., & Kemp, K. (1995). *International GIS dictionary*. John Wiley & Sons.

Merrian Webster [MW]. (2019). Kinematics. Retrieved February 12, 2019, from https://www.merriam-webster.com/dictionary/kinematics

Middel, A., Häb, K., Brazel, A. J., Martin, C. A., & Guhathakurta, S. (2014). Impact of urban form and design on mid-afternoon microclimate in Phoenix Local Climate Zones. *Landscape and Urban Planning, 122*, 16–28.

Middel, A., Chhetri, N., & Quay, R. (2015). Urban forestry and cool roofs: Assessment of heat mitigation strategies in Phoenix residential neighborhoods. *Urban Forestry & Urban Greening, 14*(1), 178–186.

Murray, A. T., & Grubesic, T. H. (2012). Spatial optimization and geographic uncertainty: Implications for sex offender management strategies. In M. Johnson (Ed.), *Community-based operations research* (pp. 121–142). New York: Springer.

Murray, A. T., Wei, R., & Grubesic, T. H. (2014). An approach for examining alternatives attributable to locational uncertainty. *Environment and Planning B: Planning and Design, 41*(1), 93–109.

Petrasova, A., Mitasova, H., Petras, V., & Jeziorska, J. (2017). Fusion of high-resolution DEMs for water flow modeling. *Open Geospatial Data, Software and Standards, 2*(1), 6.

Pix4D. (2017). *Do RTK/PPK drones give you better results than GCPs?*. Retrieved May 5, 2019, from https://tinyurl.com/yxog8vmk

Pix4D. (2019). *Pix4D mapper*. Retrieved June 27, 2019, from https://www.pix4d.com/

Robinson, A. H., Morrison, J. L., Muehrcke, P. C., Kimmerling, A. J., & Guptill, S. C. (1995). *Elements of cartography* (6th ed.). New York: Wiley.

Sensefly. (2017). eBee RTK Accuracy Assessment. Retrieved February 13, 2019, from https://tinyurl.com/ukotp2k

Tokarczyk, P., Leitão, J. P., Rieckermann, J., & Blumensaat, F. (2015). High-quality observation of surface imperviousness for urban runoff modelling using UAV imagery. *Hydrology and Earth System Sciences, 19*(10), 4215–4228.

Trammell, S. (2019). Cadastral GIS of Real Value. *Esri*. Retrieved March 12, 2019, from https://tinyurl.com/qn8o9ry

United States Census Bureau. (2019). *ZIP code tabulation areas (ZCTAs)*. Retrieved March 12, 2019, from https://tinyurl.com/y3v8ncqz

Unwin, D. J. (1995). Geographical information systems and the problem of 'error and uncertainty'. *Progress in Human Geography, 19*(4), 549–558.

Wu, J. (2006). Environmental amenities, urban sprawl, and community characteristics. *Journal of Environmental Economics and Management, 52*(2), 527–547.

Chapter 7
Capturing Neighborhood Physical Disorder Using Small, Unmanned Aerial Systems

Abstract Neighborhood physical disorder manifests in many different ways, from litter and graffiti, to abandoned cars and broken windows. The effects of physical disorder are also varied, from limiting recreational activity for both young and elderly neighborhood residents, to creating a discount effect on housing prices. Methods for capturing physical disorder in neighborhoods often include in-person observation, the use of virtual street audits, windshield tours, and satellite imagery. The purpose of this chapter is to explore how small unmanned aerial systems can be used to capture physical disorder in neighborhoods. In addition, we discuss a basic analytical framework for quantifying disorder and detecting problematic areas within the community. Limitations and implications for urban and environmental planning are discussed.

Keywords Disorder · Public space · Google Street View · Satellite · Sunnyslope · WingtraOne · Grid · Local indicators of spatial association · Planning

Neighborhood physical disorder manifests in many different ways, from litter and graffiti, to abandoned cars and broken windows. The effects of physical disorder are also varied, from limiting recreational activity for both young and elderly neighborhood residents, to creating a discount effect on housing prices. Methods for capturing physical disorder in neighborhoods often include in-person observation, the use of virtual street audits, windshield tours, and satellite imagery. The purpose of this chapter is to explore how small unmanned aerial systems can be used to capture physical disorder in neighborhoods. In addition, we discuss a basic analytical framework for quantifying disorder and detecting problematic areas within the community. Limitations and implications for urban and environmental planning are discussed.

T. H. Grubesic, J. R. Nelson, *UAVs and Urban Spatial Analysis*,
https://doi.org/10.1007/978-3-030-35865-5_7

7.1 Introduction

Although there is no single, widely accepted definition of *physical disorder* in the literature, it generally refers to the deteriorating physical condition of a neighborhood—its streets, sidewalks, building exteriors, fences, and parks (Jones et al. 2011). However, physical disorder manifests in other ways, not always directly observable from the street-level, or the naked eye. For example, the material condition of rooftops, swimming pools, tree health, and the overall conditions of semi-public spaces also contribute to perceptions of physical disorder. Grubesic et al. (2018) provide an overview of exactly what constitutes these semi-public spaces. In short, semi-public spaces consist of places like alleyways, backyards, sideyards, or other areas that may not be tied (or directly observable) to a street or block face, but nonetheless, impact residents' lives.

The literature on physical disorder is large and growing. As detailed by Jones et al. (2011), it is primarily rooted in two different disciplines—epidemiology and sociology, but with important contributions from geography, criminology, urban planning, economics, and many of the other socioeconomic planning sciences. We provide an abbreviated review of this literature in the next section. However, to set the stage for this review, it is important to acknowledge the four ways in which traditional measures of physical disorder are captured: (1) in-person observation, (2) virtual street audits, (3) windshield tours, and (4) satellite imagery. Although each of these approaches have their strengths and weaknesses, there are several compelling reasons to consider the addition of sUAS to this portfolio of neighborhood evaluation tools. The purpose of this chapter is to explore how small unmanned aerial systems (sUAS) can be used to capture physical disorder in neighborhoods, highlighting the many, powerful ways that sUAS can fill-in some of the gaps left by traditional approaches. We detail how the inclusion of semi-public spaces in physical disorder evaluations provide the geospatial intelligence necessary to better evaluate the conditions of neighborhoods. In addition, we highlight several analytical approaches for quantifying the relative severity of physical disorder in a neighborhood using sUAS imagery, a commercial geographic information system, and basic spatial statistics. The limitations of the sUAS approach are acknowledged, but we also discuss the implications that this approach may have for urban and environmental planning, public health, and law enforcement agencies.

7.2 Background

As mentioned earlier, there is a large and growing literature on physical disorder, primarily fueled by contributions from sociology and epidemiology. For example, on the public health side, Schaefer-McDaniel et al. (2010) provide a nice review and meta-analysis on the three core facets of neighborhood evaluations: (1) how the observations are carried out in the field and how the data are analyzed, (2) how neighborhoods and areas of observation are spatially defined, and (3) how the observations are ultimately linked to outcomes, in this case, health. A core finding of the Schaefer-McDaniel work is that because there is so much variability in how the observations are conducted and analyzed, it would be extremely difficult to compare findings across studies. This is not a surprising result. One only needs to consider the huge variation in how neighborhoods are defined geographically (Weiss et al. 2007; Park and Rogers 2015) to understand that the various administrative units used for this type of work (blocks, block groups, tracts, ZIP codes, city-defined neighborhoods) are not easily (or directly comparable) (Grubesic 2008). Additional substantive contributions from public health includes work that explores the impacts of physical disorder on physical activity (Molnar et al. 2004; Douglas et al. 2018), psychophysiological distress (Hill et al. 2005) and its interaction effects with child maltreatment, and binge drinking (Keyes et al. 2012), as well as childhood obesity (Franzini et al. 2009).

From a sociological and criminological perspective, it is argued that physical disorder (Sampson and Raudenbush 1999; Skogan 1990) can be an artifact of *prior* criminal behavior in the area. Specifically, if gone unchecked, the presence of physical disorder in a neighborhood can begin the process of neighborhood decay (Skogan 1990). The "spiral of decay" begins when this physical disorder fuels and intensifies additional disorder—pushing neighborhoods into a state of malaise (Ross and Mirowsky 1999), changing residents' perceptions and attitudes regarding the use of physical and social space. In turn, this affects how individuals both perceive and interact with their neighborhood. In sociology and criminology, the results of research exploring the linkages between physical disorder and crime are mixed, with some studies suggesting significant negative effects (Kelling and Coles 1997; Skogan 1990), and others finding no relationship between disorder and crime (Harcourt 2001; Sampson and Groves 1989; Sampson and Raudenbush 1999; Taylor 2001).

Grubesic et al. (2018) suggest that one of the reasons these mixed effects might exist is the use of flawed disorder measurements. Specifically, although systematic, in-person evaluations are widely viewed as the superior approach for collecting fine detailed features of physical disorder (Pliakas et al. 2017), the data collection frames used for such collection efforts are quite limited in the spaces they measure. Specifically, many important physical elements of the neighborhood remain hidden from view and completely unaccounted for, including the semi-public spaces (e.g., alleyways, backyards) detailed previously. Again, these areas are critically important to the overall physical fabric of the neighborhood, but with few exceptions (e.g., Wallace and Schalliol 2015), these areas are routinely neglected because they do not fall within measurement frames focused on the "block face." For examples of block face measures (only), see Franzini et al. (2008), Jones et al. (2011), Sampson and Raudenbush (2004)—there are many others.

Interestingly, both sociology and epidemiology recognize that elements of physical disorder measurement may be seriously flawed. For example, the work by Caughy et al. (2001) emphasizes the need to develop a brief observational method for documenting neighborhood characteristics that is easily administered but requires a minimal outlay of resources (i.e., time and money). The rapid assessment method developed by Caughy et al. (2001) roundly rejects the use of videotaped data collection (Cohen et al. 2000; Sampson and Raudenbush 1999) and windshield tours, in favor of in-person observation. Not only are the labor requirements for video-based windshield tours significant, Caughy et al. (2001, p. 226) argue that "driving any sort of vehicle through a neighborhood at a slow rate of speed for the purposes of videotaping is intrusive and disrespectful of neighborhood residents and their communities." In addition, there are likely issues of consent.

Jones et al. (2011) explore the limitations of physical disorder measurement in Los Angeles using in-person audits, combined with data from Wave 1 of the Los Angeles Family and Neighborhood Survey (L.A. FANS). Each block face in their study area was observed by multiple, trained observers, working independently, at different times. Ultimately, the results suggested that inter-rater agreement levels were high for the same block face but that the agreement levels varied considerably depending on the neighborhood elements observed. For example, the *transitory* elements of disorder (e.g., garbage, drug paraphernalia, beer, and liquor bottles) had lower levels of agreement when compared to the more *durable* elements, such as vacant lots and abandoned building (Jones et al. 2011). Results were also impacted by operational elements of the evaluation, including time of day, day of the week, and an observers' previous experience in the neighborhood.

Paquet et al. (2010) develop a new instrument for evaluating physical disorder in neighborhoods called the Montreal Neighbourhood Assessment Tool (MoNAT) and evaluate its reliability across 30 census tracts. The major contribution of this work was anchoring subjective response options to objective descriptors. For example, rather than the instrument identifying the presence of "some" trash or debris, Paquet et al. (2010) anchored these responses with specifications like, "taking up an area less than 8.5 in. × 11 in., trash not stacked." In effect, this helps remove the subjectivity of the evaluation by eliminating subjective descriptors (e.g., "few," "some," "many") from the instrument. Tests for inter-rater reliability suggest that most of the items in MoNAT (93%) had strong agreement, but measures for alcohol containers were in the low range with kappa scores less than 0.40.

Hoeben et al. (2018) advocate for a multilevel assessment of disorder, at both the street segment level and neighborhood level. The results suggest, unsurprisingly, that the reliability of the collected data improves with more observers. However, the major measurement advance from Hoeben et al. (2018) is the prescription of optimal ratios between model levels, noting that reliability scores for physical disorder will be higher when observing 14 street segments per neighborhood and allocating these segments to seven different observers. Similar results can be obtained with 20 street segments and six observers.

Mooney et al. (2014) explore the validity of a neighborhood physical disorder measurement for a virtual street audit using Google Street View (GSV) imagery. Although the internal consistency reliability was 0.93, several key limitations were noted. First, GSV imagery is limited in its temporal collection (maybe once in every 2 years), the authors had no ability to identify small indicators of physical disorder using GSV (e.g., debris), and the use of kriging generated a continuous surface but failed to account for physiographic barriers (e.g., rivers, highways).

Finally, Charreire et al. (2014) provide a solid review of the many ways remotely sensed data can be used to both define and inventory environmental characteristics for a neighborhood, from local land use to infrastructure (e.g., sidewalks, bike lanes) to elements of physical disorder (e.g., graffiti, broken windows, litter). The bulk of the studies reviewed relied upon Google Street View, but others used satellite imagery from Google Earth or Bing Maps. The results demonstrated a high level of agreement between virtual and in-person audits for durable elements (e.g., infrastructure, equipment) when compared to more transitory elements, such as aesthetics and street conditions. Charreire et al. (2014) also note that the coverage of free geospatial services such as GSV and Bing Street Side (similar to GSV) are incomplete and temporally inconsistent.

7.2.1 Semi-public Spaces and Physical Disorder

Missing from virtually all the evaluations detailed above are inventories of physical disorder in semi-public spaces. As noted previously, semi-public spaces consist of alleyways, backyards, sideyards or other urban places that may not be directly observable from the street, but nevertheless play an important role in facilitating urban services and mobility for neighborhood residents. Consider the city of Phoenix, Arizona, which has alleyways in many of its residential areas. Similar to other cities in the southwestern United States, parcel property lines are defined by privacy walls made of concrete blocks and/or fences that are approximately 6 ft high. The alleyways are considered semi-public spaces and have a range of functions for residents. Some are used for sanitation services, while others serve as the way for residents to access their garages. These alleyways have proven problematic for many neighborhoods in the city and are often described as a "highway for bad things to happen" (Janetsky 2018). Interestingly, Phoenix city code requires residents to keep the alleys behind their homes clear (Janetsky 2018), but this is difficult for both elderly residents and those without direct access to the alleys. As a result, many are unkempt, derelict, and overgrown, serving as dumping grounds for unwanted goods such as mattresses and rubbish. In some commercial and residential districts, these alleyways service as ad hoc gathering spots for the consumption of alcohol, open air drug sales, and the like. Currently, Phoenix is experimenting with a pilot program that provides gates for these alleys and limits access to them in the Sunnyslope and Royal Palm neighborhoods (Janetsky 2018).

Semi-public spaces are not just limited to alleyways. Sideyards and backyards are also considered semi-public in the state of Arizona (Fischer 2016), although this determination can vary between places in the United States. It is important to note that these areas often harbor significant elements of physical disorganization in a neighborhood, including disabled vehicles, unkempt pools, invasive, and overgrown weeds, as well as garbage and construction debris that may host zoonotic and vector-borne disease habitats (LaDeau et al. 2015; Leibler et al. 2016; Ogden 2016).

Traditional approaches for evaluating physical disorder rarely have access to these spaces. They may be obfuscated by fencing, exist in locations off the block face, or be so overgrown that it is impossible to identify important elements of disorder (debris, litter, etc.) at the surface. Furthermore, while much of the existing literature concerning the validity of methods for evaluating physical disorder is concerned with small flaws in measurement frameworks and inter-rater reliability, the glaring gap in this work is the nearly complete ignorance of semi-public spaces. This is where sUAS can help.

7.3 sUAS for Detecting Physical Disorder

One question that we receive frequently from colleagues, students, policy makers, and others regarding the use of remotely sensed data for evaluating physical disorder is, "why not just use satellite imagery from Google Earth? It's great, we use it all the time for our work." That may be true and as demonstrated above, there is certainly a large literature that makes use of these data. However, although Digital Globe (0.6 m resolution) and GeoEye (1 m resolution) imagery can be useful, it just cannot compare to the imagery generated from a quality sUAS sensor. The limitations associated with satellite imagery are illustrated later in this chapter but are also detailed elsewhere (Grubesic et al. 2018; Patino and Duque 2013). There are several important facets to reiterate. First, regardless of the application, satellite imagery is limited in spatial resolution and temporal frequency. A fundamental measure for the quality of a satellite image is ground sampling distance (GSD). GSD represents the distance between pixel centers, as measured on the ground. Larger values of GSD translate to less resolution, while smaller GSD values mean that each pixel will capture a smaller area—yielding enhanced spatial resolution. For evaluations of physical disorder, GSD is the single most important attribute of remotely sensed imagery. As detailed in Chap. 1, the best satellite imagery that money can buy is WorldView 3, which has a GSD of 31 cm. To be clear, this is excellent imagery, but it is not capable of capturing debris and other transitory measures of physical disorder. Second, the frequency of satellite image collection is also problematic. Again, as detailed in Chap. 1, WorldView can cover only a fraction of the earth's surface per day. Lastly, there are issues of cost. Satellite Imaging Corporation (GeoEye-1) charges up to $3000 per image, but only offer GSD of 46 cm (Manfreda et al. 2018). WorldView costs more.

This leaves sUAS as an extremely attractive option for capturing physical disorder in neighborhoods. Consider the image resolution provided by a WingtraOne sUAS equipped with a Sony RX1RII camera, a 42-megapixel full frame camera with a 35 mm lens. In optimal conditions, it is capable of 1 cm/pixel GSDs. To put this in perspective, the ability to capture small, transitory elements of disorder (e.g., plastic bags, bottles) is completely within the range of the WingtraOne or similarly equipped sUAS. In addition, the ability to collect imagery from both overhead and oblique (e.g., 60°) angles, from different altitudes and multiple points in time means that different perspectives are available for analysis and that longitudinal studies with different time horizons (e.g., hours, days, weeks) are a reality with sUAS. Finally, although not previously mentioned, the use of programmable sUAS with predetermined flight paths, coverage areas, altitudes, and camera angles help facilitate a rapid *entry to* and *exit from* neighborhoods for research teams. This drastically reduces exposure to potentially dangerous situations and places.

In the next section, we explore how small unmanned aerial systems can be used to capture physical disorder in a Phoenix neighborhood. We also introduce a basic analytical framework for quantifying disorder and for detecting problematic areas within the coverage area.

7.4 Capturing Physical Disorder in Sunnyslope

The city of Phoenix provides an excellent urban laboratory for exploring elements of physical disorder. Phoenix is the largest city (in terms of both land area and population) in the state of Arizona, comprising 516.7 square miles and 1.5 million people. As detailed by Grubesic et al. (2018), the general urban structure is like many other large cities in the Sunbelt, including Las Vegas, San Antonio, and Albuquerque. Thus, the approaches detailed here are generalizable to other locations.

We focus on the Sunnyslope neighborhood in Phoenix (Fig. 7.1), widely recognized as one of the most diverse demographic and socioeconomic areas of the city. Estes (2018) reports accurately that this is a neighborhood in transition. There are numerous efforts underway to attract new businesses, clean up the local streetscapes, and attract tourists to this less-frequented part of the Phoenix metropolitan area.

7.4.1 Platform and Mission Details

The mission was conducted with a WingtraOne equipped with the Sony RX1RII 42-megapixel camera. The WingtraOne is a vertical take-off and landing (VTOL) platform that weighs approximately 10 pounds with the Sony payload. It has a wingspan of 125 cm and a cruising speed of nearly 36 mph. It can achieve GSD of 0.7 cm/px, but in practice, it routinely achieves between a GSD between 1 and 2 cm/px. When using post-processing kinematics (PPK), absolute accuracies of 1 cm horizontal and 2 cm vertical can be achieved. Without PPK, the absolute accuracy is between 3 and 5 m.

For Sunnyslope, a single mission was flown in December 2018. It covered an area of 40.28 ha, or approximately 100 acres. The onboard camera collected 594 geotagged images with 70% lateral and longitudinal overlap. The average GSD for this mission was 1.36 cm.

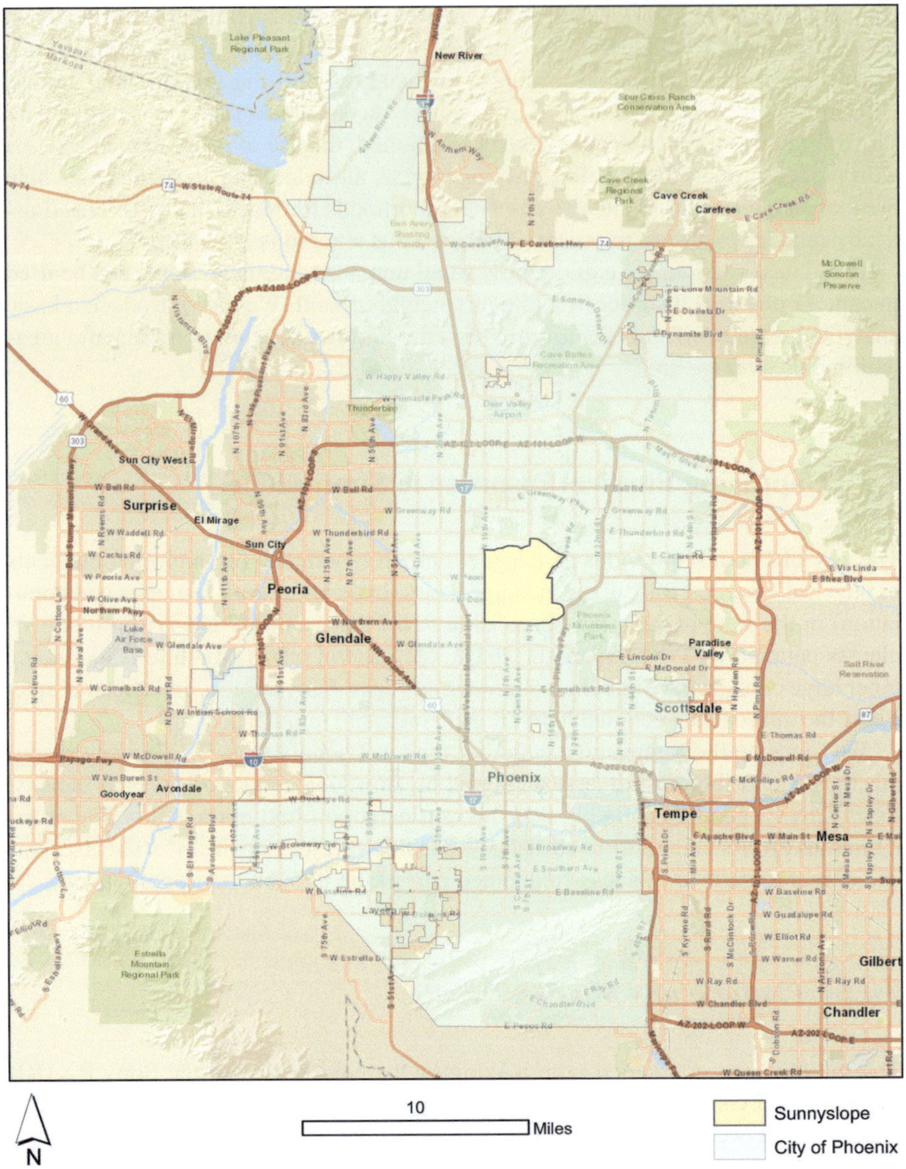

Fig. 7.1 Sunnyslope neighborhood in Phoenix, Arizona

Pix4D was used to process the acquired imagery on a windows-based personal computer with an Intel Core i7 processor at 2.8 GHz, 32 GB of RAM, and an NVIDIA Quadro K2200M graphics card. A digital orthomosaic and associated georeferenced point cloud was built for the mission. The overall computational burden for this effort was modest. Initial processing time was about 38 min. The orthomosaic required 3 h and 25 min to generate and the 3D point cloud required about 6 h of computational effort. The final point cloud consisted of 173,534,395 points with an average density of 1088.4 points per m^3.

7.4.2 Comparing sUAS Imagery with LandSat 8

One way to appreciate the quality of the WingtraOne imagery is to make a direct comparison with widely available satellite data, such as LandSat 8. Figure 7.2 displays a portion of the Sunnyslope neighborhood with LandSat 8 imagery. As detailed by Herwig (2016), both Google Earth and Google Maps include the Landsat 8 satellite imagery. Landsat 8 provides panchromatic imagery with 15 m GSD. Although the sharpness of Fig. 7.2 is not great, it is easy to identify important elements of the neighborhood physical composition—from parking lots and park play structures to individual heat pumps on the roof of many houses. To be clear, this imagery may be sufficient for many urban analyses, but compare this to Fig. 7.3 that displays the WingtraOne imagery. Although the sharpness is significantly improved, there is no dramatic increase in one's ability to identify more elements of the neighborhood's physical composition. This, of course, is an issue of spatial scale. Figure 7.4 highlights identically scaled images (1:10 m), where it becomes abundantly clear that the WingtraOne imagery far exceeds the quality of the LandSat 8. One only needs to look at the inset in Fig. 7.4 to appreciate these differences—where the address marker (1220) is clearly seen in the sUAS imagery (1:2 m). The LandSat 8 panchromatic imagery simply cannot compete.

7.4.3 Capturing Physical Disorder with sUAS Imagery

Much like the alternative approaches for capturing physical disorder, a viable assessment instrument is required for analysis. Although there is some overlap in the types of disorder elements one would be examining with sUAS imagery and with in-person evaluations (see Paquet et al. 2010 for an example) because of the bird's eye view afforded by the sUAS, it demands some unique categories. As detailed previously, semi-public spaces may include backyards, sideyards, and various elements one might find in those locations (e.g., swimming pools). One can also explore elements that may be obfuscated from street-level observations, including things like roof damage or the debris found in limited-access alleyways.

For the purposes of this chapter, we are going to assume that readers interested in this type of work have developed their own, comprehensive instrument, or at the very least have access to one developed by others (which can then be tested and modified). Table 7.1 represents a *very* simplistic instrument for tabulating physical disorder cues in remotely sensed imagery. However, this is not something that we recommend using for a full-blown study on neighborhood disorder. In this instance,

Fig. 7.2 LandSat8 imagery for Norton Park in the Sunnyslope neighborhood

we are attempting to keep the instrument as modest as possible, for illustrative purposes.

Once the chosen instrument is ready for use, the process of coding disorder with the sUAS imagery is straightforward. For the purposes of this analysis, a 10 m × 10 m vector grid is structured over the study area of interest. Many types of geographic

Fig. 7.3 sUAS-derived imagery for Norton Park in the Sunnyslope neighborhood

tessellations would work for this (e.g., triangles, hexagons), but whatever the choice, the cells need to cover the entire study area without underlap or overlap. As illustrated in Fig. 7.5, a circle is not a good option because when spaced evenly, there are large gaps in coverage of the study area. Further, if packed more tightly, there would be overlap. In addition, it is important to acknowledge that the size of the grid cells

Fig. 7.4 Comparing LandSat 8 to sUAS-derived imagery

chosen for analysis will impact one's results. For example, if we were to use 5 m × 5 m or 20 m × 20 m cells, the way in which disorder cues are aggregated to each cell will be dramatically different, as well as the resulting statistical analyses on disorder. This is known as the modifiable areal unit problem (MAUP) and is a

Table 7.1 Physical disorder instrument

Grid cell #	Litter	Liquor bottles	Roof damage	Derelict pool	Construction debris	Large debris
1	1	0	0	0	0	0
2	1	0	0	0	0	0
3	0	0	0	0	0	0
4	0	0	0	0	0	0
5	0	0	0	0	0	0
6	0	0	0	0	0	0
7	0	0	0	0	0	0
8	0	0	0	0	0	0
…	…	…	…	…	…	…
380	1	0	0	0	0	0

For each cue, write down the number of times it appears in the grid cell

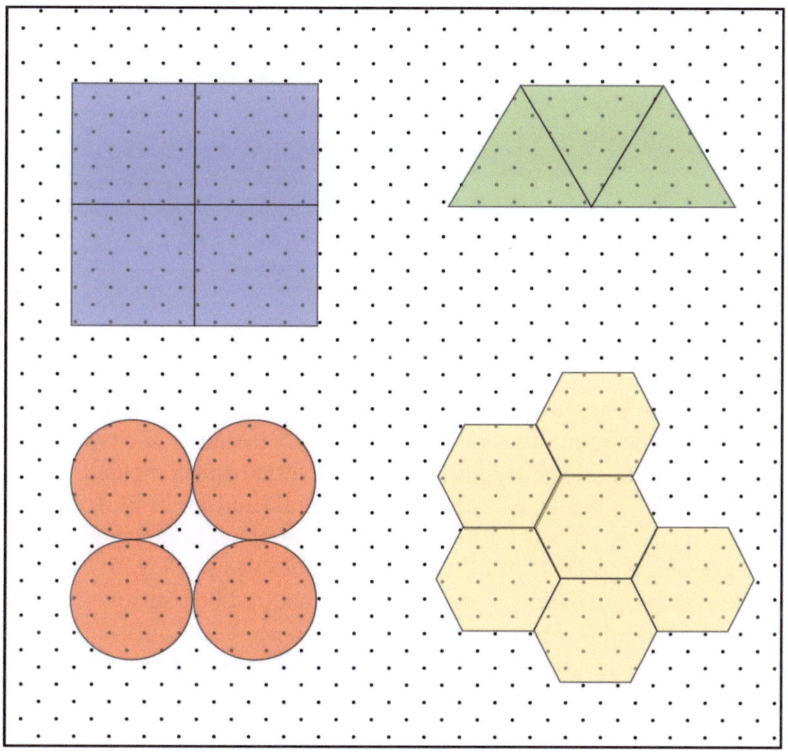

Fig. 7.5 Geometric tessellations for evaluating physical disorder

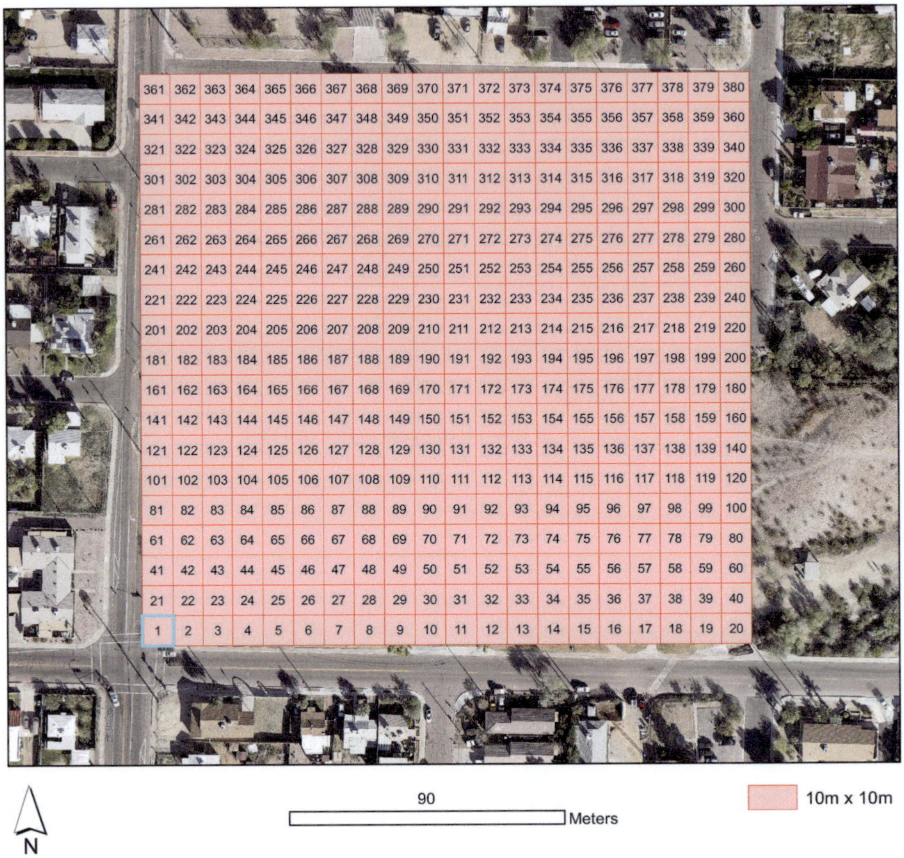

Fig. 7.6 10 m × 10 m grid for evaluating physical disorder

widely recognized limitation in spatial analysis (Openshaw 1984; Jelinski and Wu 1996). Efforts to ensure rigorous results would likely include some type of sensitivity analysis, where different sized grid cells were tested and compared. Each of the grid cells ($n = 380$) in the study area have a unique ID and are displayed in Fig. 7.6. The process of coding physical disorder basically amounts to having multiple raters evaluate each of the grid cells for disorder cues. For example, Fig. 7.7 displays cell #1. For illustrative purposes, we have added a small red box surrounding the only disorder cue in the cell—a piece of litter near the curb. These cues can be added to Table 7.1 as an entry in a flat file database or added directly into the spatial database in a commercial GIS program. Figure 7.8 illustrates a significantly more complex landscape of disorder cues. In this instance, a box spring mattress (large debris) is found in the alley, along with significant amounts of litter and liquor bottles, as well as some litter and larger debris in a backyard. Raters for this cell would have the opportunity to "zoom in" to identify the exact counts of these disorder cues for entry into Table 7.1. Once all 380 cells are evaluated by Rater 1, the entire process is

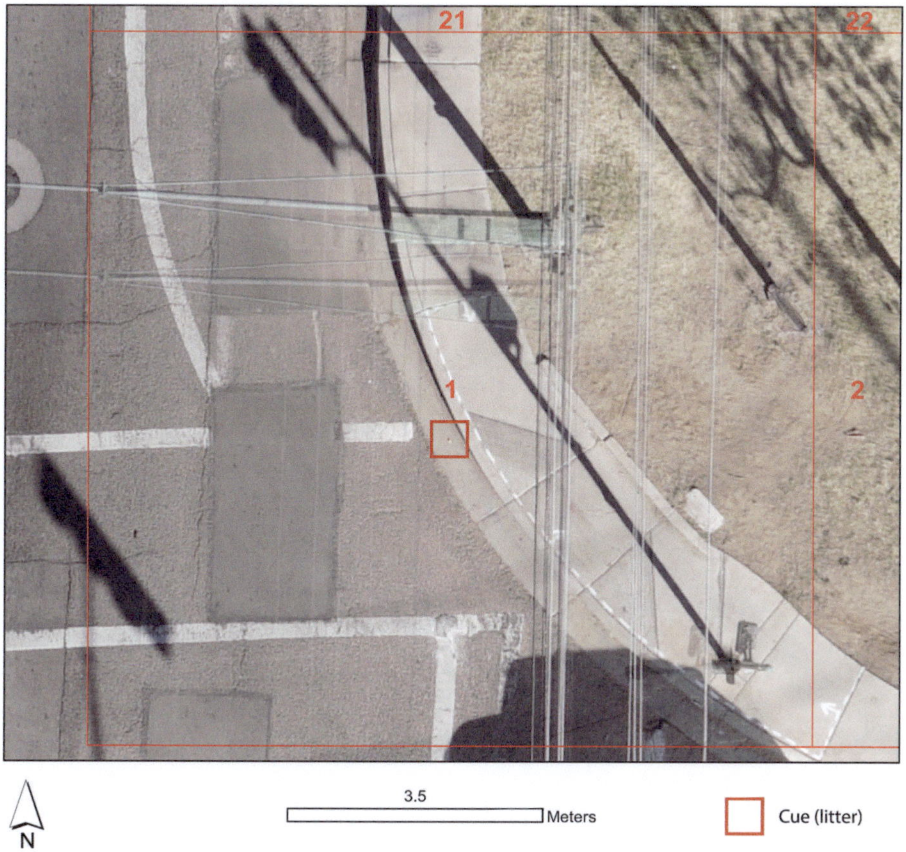

Fig. 7.7 Operationalizing the grid for evaluating physical disorder

repeated by Raters 2, 3, 4,…, *N*. When the process is complete, inter-rater reliability (Hallgren 2012) can be calculated, and the validity of the instrument can be evaluated (Hoeben et al. 2018; Molnar et al. 2004; Pliakas et al. 2017).

7.4.4 Measuring Physical Disorder with sUAS Imagery

Once the tabulations are complete, there are a variety of options for exploring the patterns of physical disorder, as well as connecting these patterns to process. One way to approach this is through geovisualization (Dykes et al. 2005). Figure 7.9 highlights the overall landscape of disorder for the neighborhood using a graduated symbol map (Brewer and Campbell 1998). Graduated symbol maps help identify the varying levels of disorder in each grid cell and are the appropriate choice for count data. There appear to be several problematic areas north of the park, most

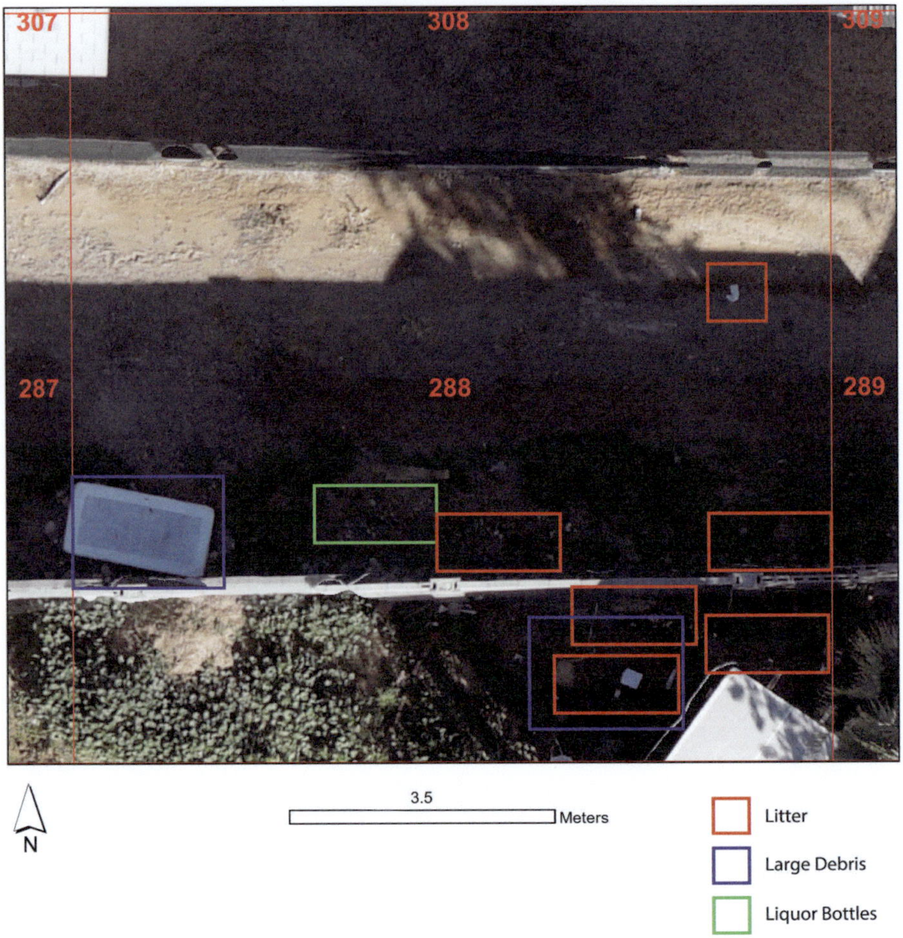

Fig. 7.8 A more complex landscape of physical disorder cues

notably in the alley that splits the northern housing block. Recall Fig. 7.8, where a mattress, several piles of debris and multiple alcohol bottles were identified. Figure 7.9 suggests that the interior portions of the alley (approximately 30–40 m away from the major cross streets) function as a preferred dumping ground. Astute readers will also note that some of the homes in this area display significant signs of physical disorder on the roof, including missing shingles, poor seam repairs, mismatched materials, damaged HVAC pipes, and the like. Another interesting result displayed by Fig. 7.9 is the relative lack of disorder in the park. Aside from the parking lot, there were only a few stray disorder cues, primarily litter (e.g., wrappers, plastic waste) scattered throughout.

One way to capture the relative differences in disorder through the study area is through a standard deviation map. The standard deviation map, while relatively simple, is a power tool for disambiguating levels of disorder throughout a neighbor-

Fig. 7.9 Representing physical disorder: total counts of disorder cues, by cell

hood. So, while it may be painful, it is time for readers to recall material from their introductory statistics class. The standard deviation for a variable (e.g., count of physical disorder cues) representing non-grouped data is calculated using the following formula:

$$\sigma = \sqrt{\frac{1}{N}\sum_{i=1}^{N}\left(x_i - \mu\right)^2} \tag{7.1}$$

where x_i is the count of the physical disorder cues for a given grid cell and μ is the mean of all values in the study area. Once σ is calculated, each standard deviation becomes a class for the choropleth map. For example, one could use a value of 1 SD for each class, 0.50 SD, 0.25 SD, and so on. For the study area in Sunnyslope, the

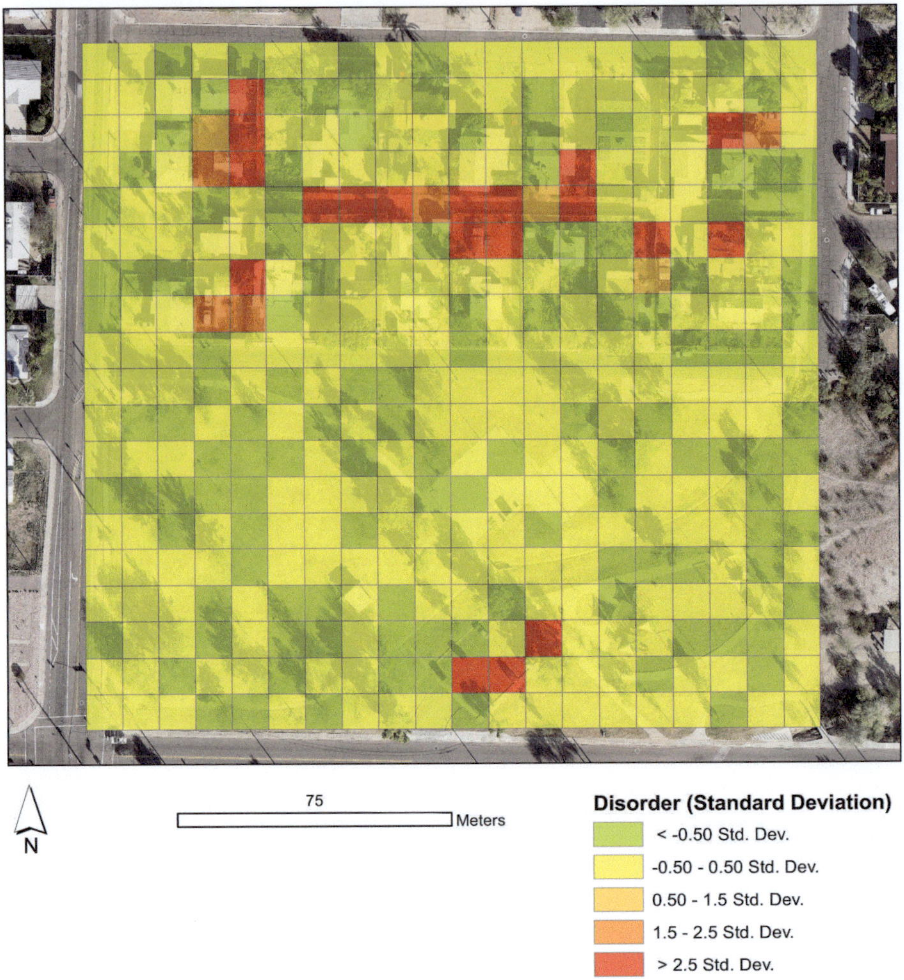

Fig. 7.10 Standard deviation of physical disorder cues in Sunnyslope

mean value is 1.82, while the minimum value is 0, the maximum is 20, and the standard deviation is 3.46. Figure 7.10 displays the associated standard deviation map. Notice that *most* of the values for the study area are within 0.50 standard deviation of the mean or lower. This is notable because it suggests that the relative levels of physical disorder for the neighborhood are similar. However, there are several cells with values higher than the mean (>0.50), which suggests that portions of the neighborhood have elevated levels of physical disorder. Again, the alley north of the park, several of the houses in that same block, as well as the park's parking lot appear to be problematic.

Finally, readers may want to consider using more rigorous statistical methods for identifying statistically significant hot spots of disorder. This would be especially useful when exploring the geographic intricacies of a larger study area. One way to approach this is to develop a basic spatial taxonomy of physical disorder. Local indicators of spatial association (LISA), particularly the local Moran's I measure, are an effective way to accomplish this (Anselin 1995). At their core, LISA statistics identify outliers—places that display a statistically significant higher or lower levels of a measured outcome. These outcomes are the variable of interest, which in this case is physical disorder. However, it could be anything, ranging from crime and disease to income and poverty. What makes LISA approaches special is their use of spatial relationships to identify geographic variations in the variable of interest. Thus, rather than simply treating each grid cell, i, as an independent observation, completely isolated from other grid cells, j, LISA uses spatial weighting procedures to account for the influence of spatially proximal cells. There are a variety of different strategies for establishing spatial weights, from contiguity (e.g., rook or queen) relationships, to distance (network or Euclidean) to k-nearest neighbors. For more details on the development of spatial weights matrices and their implications for geographic analysis, see Getis and Ord (1996) or Anselin and Rey (2014).

Consider the following notation (Anselin 2002):

$z_k = \dfrac{[x_k - \overline{x}_k]}{\sigma_k}$ or a standardized random variable with a mean equal to zero and astandard deviation equal to 1.

$z_l = \dfrac{[x_l - \overline{x}_l]}{\sigma_l}$ or a standardized random variable with a mean equal to zero and astandard deviation equal to 1.

W is typically a Euclidean (straight line) row standardized spatial weights matrix with binary values of 0 or 1. However, W can also be specified using a simple spatial adjacency metric such as queen's contiguity, where both edges and vertices of the administrative unit are evaluated for proximity. n is the number of observations. Given this basic notation, the global Moran's I statistic is specified as follows:

$$I = \frac{n\sum_k \sum_l W_{z_k z_l}}{W\sum_k z_k^2} \tag{7.2}$$

The local Moran's I is specified as:

$$I_k = z_k \sum_l W z_l \tag{7.3}$$

The spatial weights are typically row-standardized, so it is not necessary to account for the usual scaling factors since $S_0 = \sum_i \sum_j w_{ij} = n$ and thus $\left(\frac{n}{S_0}\right)(z_k^{'} W z_l \,/\, z_k^{'} zk) = z_k^{'} W z_l \,/\, z_k^{'} zk$.

For the purposes of this analysis, we utilized a queen's contiguity spatial weights matrix and the local Moran's I test for detecting spatial association. The Moran's I

value was 0.201, which is not particularly high, but it was statistically significant and suggests that spatial autocorrelation is present. Figure 7.11 displays the patterns of spatial association for physical disorder between cells. From an interpretive standpoint, high-high (red, $n = 24$) corresponds to cells with high levels of physical disorder surrounded by other cells with similar levels of disorder. Low-low (blue, $n = 21$) corresponds to cells with low levels of physical disorder surrounded by cells with similar levels of disorder. The low-high (light blue, $n = 16$) category represents locations with low levels of physical disorder surrounded by cells with relatively higher levels of disorder. Finally, the high-low (pink, $n = 7$) category corresponds to locations with high levels of physical disorder surrounded by cells with relatively lower levels of disorder.

There are several interesting aspects of Fig. 7.11 worth noting. The large grassy space in the park west of the basketball courts and play structure display very low levels of physical disorder. However, there is a high-low area proximal to the picnic shelter in the park that is an interesting anomaly. There was a variety of litter scattered about this area, which is not surprising given the level of use the area receives. Also, notice the parking lot (southwest of the play structure) and its low-high LISA values. Recall the standard deviation map (Fig. 7.10)—which highlighted the parking lot as having relatively high levels of disorder for the study area. However, only one grid cell is included as a hot spot in Fig. 7.11. This highlights one of the more powerful aspects of the LISA tests. Although there is physical disorder present in the parking lot, it is not sufficiently high enough to generate a larger, statistically significant hot spot. Remember how the LISA works—it incorporates values from the local neighborhood (j) and provides a spatially smoothed value for the cell under evaluation (i) for statistical testing. Here, because most of the parking lot cells were free of disorder, only one cell was considered a hot spot. The LISA identified a handful of really low values (light blue) proximal to some cells with relatively higher values. In short, this is why multiple geovisualization and analytical approaches are required to develop a complete understanding of physical disorder for a neighborhood.

As one might have guessed, the most problematic locations for the study area are located north of the park. Not only is the alley that splits this block loaded with physical disorder, many of the homes in this area are also showing signs of disrepair and disorder. These are semi-public spaces, and most would be completely missed using traditional neighborhood evaluation techniques. For example, this chapter already established that publicly available satellite imagery does not have the resolution to pick up the more transitory elements of physical disorder. Beyond that, it is important to note that GSV would have almost no information on the problematic alley. As detailed by Grubesic et al. 2018, although disorder cues near the entrances to the alley (on both sides of the block) would be captured, the major physical disorder is located at least 30 m down the alley, far out-of-range of the GSV cameras. Could an in-person audit or windshield tour capture this information? Yes, but only some of it. The problematic rooftops, as well as the debris behind walls and fences would be missed. Where the analytical methods are concerned, at the very least, Fig. 7.11 reinforces the patterns evident in Figs. 7.9 and 7.10, but does so in a more rigorous, *spatially informed* way.

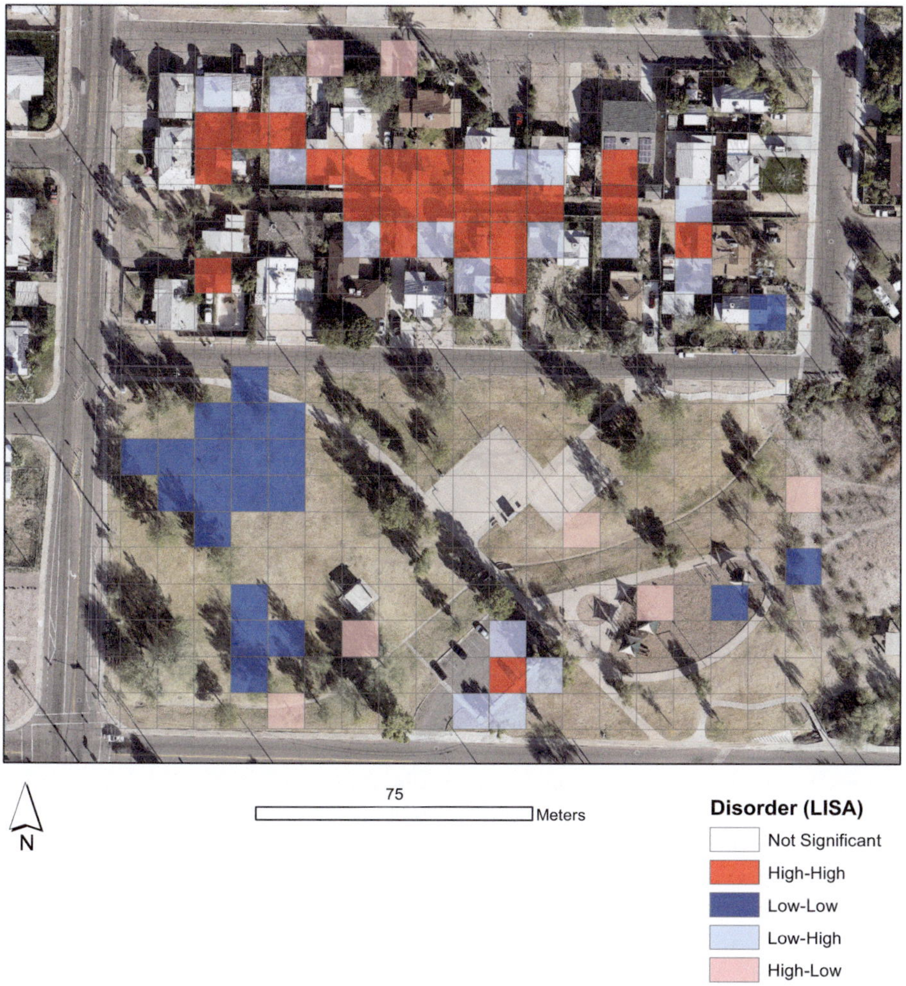

Fig. 7.11 Identifying spatial clusters of physical disorder using local indicators of spatial association

7.5 Limitations and Implications for Urban and Environmental Planning

Given the results presented above, what does it mean for urban and environmental planning and the other disciplines that focus on physical disorder measurements? There are three major takeaways. First, the use of sUAS for remotely sensing physical disorder provides a massively different perspective on the problem, both literally and figuratively. Can it replace in-person audits? No. Can it radically compliment such efforts? Yes. When conducted with care and a significant investment of time, in-person audits provide a resolution not available to sUAS. The ability to identify condoms, cigarette butts and micro-elements of transitory disorder are much easier, in person. While the imagery for sUAS is certainly high-quality, the ability to identify these small items remains somewhat elusive. Further, because the sUAS is operating from above, there is no easy way to evaluate building facades, the quality of their siding, bars on windows or many of the physical disorder cues that require a strongly oblique (or perpendicular) viewpoint. Second, the results of this analysis suggest that the use of sUAS for identifying physical disorder in *semi-public* spaces is unmatched by any traditional neighborhood audit approach. Again, the high-resolution access to rooftops, backyards, sideyards, and alleyways provides a tremendous amount of data and information that cannot be collected in any other way, at least without risking life and limb, or trespassing. In addition, the temporal frequency in which these data can be collected is an advantage. Although it certainly takes time to code and evaluate the imagery, the collection of these data requires minimal effort. For example, although we focused only on a subset of our study area in Sunnyslope, the WingtraOne was able to capture nearly 600 images covering 100 acres in about 15 min of flight time. This scale of data collection, even for a large team, would require multiple days of field work for an in-person neighborhood audit. Third, because the sUAS data are spatially explicit, they lend themselves to a range of rigorous analytical techniques that are not as easily applied with data generated from more traditional approaches. For example, it would be trivial to calculate the size of the mattress found in the alley or to calculate the percentage of a roof that is damaged on a given house. Similarly, identifying spatial relationships at a granular level (e.g., distance, proximity, clustering) becomes easier with sUAS-derived imagery. As detailed in this chapter, the use of basic geovisualization techniques and more advanced spatial statistics is easily done with sUAS data and a geographic information system. Hidden elements of physical disorder including pool water quality (e.g., emerging algae bloom) and tree/plant health could also be identified in a spatially explicit way with a multispectral sensor mounted to the sUAS.

In sum, sUAS excel in collecting high-resolution data on physical disorder for neighborhoods. When coupled with a geographic information system for visualization and analysis, systematic patterns of disorder can be detected. sUAS cannot and should not be viewed as a replacement for structured, in-person field work; however, sUAS data add a level of richness and detail that cannot be achieved using traditional approaches. Beyond physical disorder evaluations for scientific inquiry, sUAS can offer cities a powerful alternative to large-scale urban infrastructure assessments that currently require large outlays of human resources. From evaluating the quality of street and sidewalk infrastructure, to the health of urban vegetation, the strategic deployment of sUAS throughout a city can save both time and money. Further, because the aerial audits can be routinized, it minimizes the chances for accidentally missing a street, alleyway, or park due to unintended human error. The use of these data also eliminates elements of subjectivity when evaluating urban amenities and infrastructure. For example, how can one be sure that a patch of pavement is distressed? It may look distressed to some people, but not others. Image-based three-dimensional reconstruction of street surfaces using sUAS datasets can help address these uncertainties (Inzerillo et al. 2018). There is little doubt that from the perspective of planning and management, the ability to reduce data uncertainty is important for ensuring equitable infrastructure investments (e.g., handicap accessible sidewalks, urban greening programs). sUAS can help cities do this with higher accuracy, efficiency, and safety, over time and space.

References

Anselin, L. (2002). Under the hood Issues in the specification and interpretation of spatial regression models. *Agricultural Economics, 27*(3):247–267.

Anselin, L. (1995). Local indicators of spatial association—LISA. *Geographical Analysis, 27*(2), 93–115.

Anselin, L., & Rey, S. (2014). *Modern spatial econometrics in practice: A guide to GeoDa, GeoDaSpace and PySAL*. Chicago: GeoDa Press.

Brewer, C., & Campbell, A. J. (1998). Beyond graduated circles: Varied point symbols for representing quantitative data on maps. *Cartographic Perspectives,* (29):6–25.

Caughy, M. O., O'Campo, P. J., & Patterson, J. (2001). A brief observational measure for urban neighborhoods. *Health & Place, 7*(3), 225–236.

Charreire, H., Mackenbach, J. D., Ouasti, M., Lakerveld, J., Compernolle, S., Ben-Rebah, M., & Oppert, J. M. (2014). Using remote sensing to define environmental characteristics related to physical activity and dietary behaviours: A systematic review (the SPOTLIGHT project). *Health & Place, 25*, 1–9.

Cohen, D., Spear, S., Scribner, R., Kissinger, P., Mason, K., & Wildgen, J. (2000). " Broken windows" and the risk of gonorrhea. *American Journal of Public Health, 90*(2), 230.

Douglas, J. A., Briones, M. D., Bauer, E. Z., Trujillo, M., Lopez, M., & Subica, A. M. (2018). Social and environmental determinants of physical activity in urban parks: Testing a neighborhood disorder model. *Preventive Medicine, 109*, 119–124.

Dykes, J., MacEachren, A. M., & Kraak, M. J. (2005). *Exploring geovisualization*. Amsterdam: Elsevier.

Estes, C. (2018). Can phoenix's sunnyslope community become a tourist destination? *KJZZ*. Retrieved March 14, 2019, from https://tinyurl.com/ybcgs5j6

Fischer, H. (2016). New law allows drones to photograph people basking in their back yards. *Arizona Capitol Times*. Retrieved April 24, 2018, from https://tinyurl.com/y7vvxbnq

Franzini, L., Caughy, M. O. B., Nettles, S. M., & O'Campo, P. (2008). Perceptions of disorder: Contributions of neighborhood characteristics to subjective perceptions of disorder. *Journal of Environmental Psychology, 28*(1), 83–93.

Franzini, L., Elliott, M. N., Cuccaro, P., Schuster, M., Gilliland, M. J., Grunbaum, J. A., & Tortolero, S. R. (2009). Influences of physical and social neighborhood environments on children's physical activity and obesity. *American Journal of Public Health, 99*(2), 271–278.

Getis, A., & Ord, J. K. (1996). Local spatial statistics: An overview. In *Spatial analysis: Modelling in a GIS environment* (Vol. 374, pp. 261–277). Bristol: GeoInformation International.

Grubesic, T. H. (2008). Zip codes and spatial analysis: Problems and prospects. *Socio-Economic Planning Sciences, 42*(2), 129–149.

Grubesic, T. H., Wallace, D., Chamberlain, A. W., & Nelson, J. R. (2018). Using unmanned aerial systems (UAS) for remotely sensing physical disorder in neighborhoods. *Landscape and Urban Planning, 169*, 148–159.

Hallgren, K. A. (2012). Computing inter-rater reliability for observational data: An overview and tutorial. *Tutorials in Quantitative Methods for Psychology, 8*(1), 23.

Harcourt, B.E. (2001). *Illusion of Order. The False Promise of Broken Windows Policing*. Harvard University Press, Cambridge, MA.

Herwig, C. (2016). Keeping Earth up to date and looking great. *Googleblog*. Retrieved May 16, 2018, from https://tinyurl.com/y9gj3dqb

Hill, T. D., Ross, C. E., & Angel, R. J. (2005). Neighborhood disorder, psychophysiological distress, and health. *Journal of Health and Social Behavior, 46*(2), 170–186.

Hoeben, E. M., Steenbeek, W., & Pauwels, L. J. (2018). Measuring disorder: Observer bias in systematic social observations at streets and neighborhoods. *Journal of Quantitative Criminology, 34*(1), 221–249.

Inzerillo, L., Di Mino, G., & Roberts, R. (2018). Image-based 3D reconstruction using traditional and UAV datasets for analysis of road pavement distress. *Automation in Construction, 96*, 457–469.

Janetsky, M. (2018). 'Highway for bad things': Phoenix tries closing alleys after neighbors complain of crime. *Arizona Republic*. Retrieved from https://tinyurl.com/y8tor8cl

Jelinski, D. E., & Wu, J. (1996). The modifiable areal unit problem and implications for landscape ecology. *Landscape Ecology, 11*(3), 129–140.

Jones, M., Pebley, A. R., & Sastry, N. (2011). Eyes on the block: Measuring urban physical disorder through in-person observation. *Social Science Research, 40*(2), 523–537.

Kelling, G. L., & Coles, C. M. (1997). *Fixing broken windows: Restoring order and reducing crime in our communities*. Simon and Schuster.

Keyes, K. M., McLaughlin, K. A., Koenen, K. C., Goldmann, E., Uddin, M., & Galea, S. (2012). Child maltreatment increases sensitivity to adverse social contexts: Neighborhood physical disorder and incident binge drinking in Detroit. *Drug and Alcohol Dependence, 122*(1–2), 77–85.

LaDeau, S. L., Allan, B. F., Leisnham, P. T., & Levy, M. Z. (2015). The ecological foundations of transmission potential and vector-borne disease in urban landscapes. *Functional Ecology, 29*(7), 889–901.

Leibler, J. H., Zakhour, C. M., Gadhoke, P., & Gaeta, J. M. (2016). Zoonotic and vector-borne infections among urban homeless and marginalized people in the United States and Europe, 1990–2014. *Vector-Borne and Zoonotic Diseases, 16*(7), 435–444.

Manfreda, S., McCabe, M. F., Miller, P. E., Lucas, R., Pajuelo Madrigal, V., Mallinis, G., & Müllerová, J. (2018). On the use of unmanned aerial systems for environmental monitoring. *Remote sensing, 10*(4), 641.

Molnar, B. E., Gortmaker, S. L., Bull, F. C., & Buka, S. L. (2004). Unsafe to play? Neighborhood disorder and lack of safety predict reduced physical activity among urban children and adolescents. *American Journal of Health Promotion, 18*(5), 378–386.

Mooney, S. J., Bader, M. D., Lovasi, G. S., Neckerman, K. M., Teitler, J. O., & Rundle, A. G. (2014). Validity of an ecometric neighborhood physical disorder measure constructed by virtual street audit. *American Journal of Epidemiology, 180*(6), 626–635.

Ogden, N. H. (2016). Emerging challenges of vector-borne diseases and cities: Vector-borne disease, climate change and urban design. *Canada Communicable Disease Report, 42*(10), 202.

Openshaw, S. (1984). Ecological fallacies and the analysis of areal census data. *Environment and Planning A, 16*(1), 17–31.

Paquet, C., Cargo, M., Kestens, Y., & Daniel, M. (2010). Reliability of an instrument for direct observation of urban neighbourhoods. *Landscape and Urban Planning, 97*(3), 194–201.

Park, Y., & Rogers, G. O. (2015). Neighborhood planning theory, guidelines, and research: Can area, population, and boundary guide conceptual framing? *Journal of Planning Literature, 30*(1), 18–36.

Patino, J. E., & Duque, J. C. (2013). A review of regional science applications of satellite remote sensing in urban settings. *Computers, Environment and Urban Systems, 37*, 1–17.

Pliakas, T., Hawkesworth, S., Silverwood, R. J., Nanchahal, K., Grundy, C., Armstrong, B., et al. (2017). Optimising measurement of health-related characteristics of the built environment: Comparing data collected by foot-based street audits, virtual street audits and routine secondary data sources. *Health & Place, 43*, 75–84.

Ross, C. E., & Mirowsky, J. (1999). Disorder and decay the concept and measurement of perceived neighborhood disorder. *Urban Affairs Review, 34*(3), 412–432.

Sampson, R. J., & Groves, W. B. (1989). Community structure and crime: Testing social-disorganization theory. *American Journal of Sociology, 94*(4), 774–802.
Sampson, R. J., & Raudenbush, S. W. (1999). Systematic social observation of public spaces: A new look at disorder in urban neighborhoods. *American Journal of Sociology, 105*(3), 603–651.
Sampson, R. J., & Raudenbush, S. W. (2004). Seeing disorder: Neighborhood stigma and the social construction of "broken windows". *Social Psychology Quarterly, 67*(4), 319–342.
Schaefer-McDaniel, N., Caughy, M. O. B., O'Campo, P., & Gearey, W. (2010). Examining methodological details of neighbourhood observations and the relationship to health: A literature review. *Social Science & Medicine, 70*(2), 277–292.
Skogan, W. G. (1990). *Disorder and decline: Crime and the spiral of decay in American neighborhoods*. Berkeley: University of California Press.
Taylor, R. B. (2001). *Breaking away from broken windows: Evidence from Baltimore neighborhoods and the nationwide fight against crime, grime, fear and decline*. New York, NY: Westview.
Wallace, D., & Schalliol, D. (2015). *Testing the temporal nature of social disorder through abandoned buildings and interstitial spaces. Social science research, 54*, 177–194.
Weiss, L., Ompad, D., Galea, S., & Vlahov, D. (2007). Defining neighborhood boundaries for urban health research. *American Journal of Preventive Medicine, 32*(6), 154–S159.

Chapter 8
Unmanned Aerial Systems for Energy Infrastructure Assessment

Abstract As access to remotely sensed imagery increases, researchers and practitioners continue to leverage these data for improving models of the built environment. Recent work illustrates the benefits of remotely sensed data, including their flexibility for extracting urban features such as buildings, vegetation, critical infrastructure, and their use in creating digital surface models. This combination of data and modeling provides important geospatial intelligence to urban planners, engineers, and other stakeholders who are looking to build smarter, more efficient, and sustainable cities. As calls for the use of clean, renewable energy increases, the ability to locate, build, maintain, and evaluate energy infrastructure grows. The purpose of this chapter is to provide an overview of how small unmanned aerial systems (sUAS) provide a low-cost and flexible way to collect high-resolution data for energy planning applications. A case study for residential solar energy audits in Phoenix, Arizona highlights the benefits of using sUAS in energy planning, infrastructure development, and maintenance.

Keywords Energy · Solar · Inspection · Point cloud · Radiation · Slope · Aspect

8.1 Introduction

Unmanned aerial systems exhibit several unique characteristics that provide them with advantages for monitoring infrastructure systems. For example, sUAS come in a variety of sizes, and their variations in design (see Chap. 2) provide different methods for capturing geospatial data. Multi-rotor systems allow for vertical takeoff and landing and have the capability to hover in place. These systems are particularly useful for inspecting wind turbines, powerlines, and poles, or other relatively compact and stationary energy infrastructure. For other infrastructure, it may be more advantageous to use fixed wing systems. As discussed in Chap. 2, fixed wing systems can fly longer distances, covering more ground. Many of these systems have a downward facing camera, and the images these systems collect can be post-processed to create high-resolution orthophotos. With this approach, entire networks of energy-related infrastructure (such as pipelines or power lines) can be

T. H. Grubesic, J. R. Nelson, *UAVs and Urban Spatial Analysis*,
https://doi.org/10.1007/978-3-030-35865-5_8

mapped and monitored at very fine scales and at routine time intervals. An additional benefit of capturing high-resolution infrastructure data is the ability to construct 3D models that can be used to make volumetric calculations, surface area assessments, monitor construction progress, or identify inefficient system components (Ham et al. 2016).

There are several active areas of research aimed at improving the spatial and temporal resolution of sUAS infrastructure data including the development of "swarm" technology (Vian et al. 2011). As the name implies, swarm technology allows multiple unmanned aerial vehicles (UAVs) to operate in conjunction with one another. They communicate using advanced sensing equipment and are deployed with the intent of capturing more data in less time (Hoffmann et al. 2011; Oung and D'Andrea 2011). However, scholars have pointed out that these and other advances have yet to be widely applied outside laboratory settings or controlled outdoor environments (Ham et al. 2016). In practice, the majority of energy-related sUAS applications are performed with a single unit (Kanellakis et al. 2019). This includes sUAS deployment for pre-installation assessments, monitoring during construction or maintenance, and continued monitoring for preventative maintenance and planning (Ham et al. 2016; Rakha and Gorodetsky 2018).

Despite limitations in the number of vehicles that can be used simultaneously, sUAS provides important information to planners, managers, and engineers that would otherwise take significantly more time and resources to collect. With increasing interest in moving city energy portfolios to more sustainable and cleaner sources, stakeholders are interested in developing more energy efficient "smart" or "green" infrastructure. In particular, cities and building engineers have a growing interest in conducting energy audits for individual buildings (Rakha and Gorodetsky 2018). The goal of such efforts is to quickly identify how and where buildings are performing at less than optimal standards. sUAS equipped with thermal sensors are being used to locate air leaks, rust, and deterioration that contribute to losses in efficiency (Mason 2018).

Another major application area for sUAS on the energy efficiency front is in calculating solar energy potential of urban areas (Redweik et al. 2013). In dense urban environments, some of the most convenient and efficient locations for solar panel installation include building roof tops and facades. A solar audit provides planners with information on how much solar could be generated in a city based on the orientation, surface area, and slope of building roof tops. One way to accomplish solar audits more efficiently is by using sUAS technology. While existing strategies rely (almost exclusively) upon light detection and ranging (LiDAR) data (Freitas et al. 2015), there is room for improvement. Namely, LiDAR data collection is temporally limited. Flights are infrequent, expensive, and only provide a single snapshot in time. sUAS can fly regularly, keeping pace with dynamic urban environments. For example, would the solar potential of a city pre/post hurricane be identical? Likely not. Of course, flying a new LiDAR mission for data collection post-hurricane is always an option, but this requires significant planning, lead-time, and a substantial outlay of resources. sUAS can be deployed immediately and frequently, targeting particular communities and neighborhoods. In short, the flexibility of sUAS is unmatched. But, to be clear, we are not suggesting that LiDAR-based assessments are obsolete. On the contrary, we are suggesting that both the temporal and spatial resolution of these assessments can be enhanced by performing them with high-resolution sUAS sensors.

Although the use of sUAS for energy-related assessments is still in its infancy (relatively speaking), it is quickly growing (Ham et al. 2016). There are many companies that are offering inspection services for energy structures, including wind turbines, oil-related facilities, and power line monitoring. Furthermore, several of the bigger UAS companies are designing systems specifically for inspection services. As a result, it is important to take stock and provide readers with an overview of the current literature focusing on the application of sUAS to energy planning. In particular, we believe that sUAS will be highly useful in the planning and development of residential solar power. This motivates us to provide a basic methodological outline for conducting solar power installation assessments in an urban environment using an sUAS.

8.2 Close Inspectors of the Drone Kind

A growing subsector of the drone industry is dedicated to developing platforms for the sole purpose of inspecting, analyzing, and mapping energy infrastructure. The low costs of operating sUAS combined with reductions in risk to human resources when inspecting energy systems represent two of the most attractive benefits for considering drone-based workflows. Difficult-to-reach areas, including transmission lines or wind turbine blades, are much more accessible with an sUAS. Further, work that might require a human several hours of walking/assessment along energy infrastructure corridors can be captured in less than an hour with an sUAS. Another benefit of using sUAS for inspection work is community risk reduction. For example, when electrical substations experience errors, protocol dictates they be taken offline for a human worker to inspect them. This may not be necessary when using sUAS technology. Specifically, prior to the allocation of human resources, an sUAS can be deployed to identify and/or troubleshoot the problem. If the situation can be resolved remotely, without dispatching a repair crew or disrupting power to a neighborhood, this is a win—reducing down-time. In other cases, errors may require the dispatch of human resources, but teams can use the information collected by the sUAS to gain a better understanding of error context, prior to arriving on-site. Again, this is a win because teams are not blindly reporting to what could be a dangerous situation in the field.

Given these advantages, it is not surprising that companies are beginning to leverage sUAS technology for such purposes. For instance, Measure is a company with offices in Colorado, Illinois, and Ohio and offer sUAS inspection services for energy infrastructure. Their website (https://www.measure.com/) claims that solar farm inspection by sUAS can cost 46% less than a manual inspection. It is reported that sUAS inspections take about 10 min per megawatt (MW) of installed solar, which saves an estimated $1200/MW inspected, on average. One reason for this high level of efficiency is the ability of sUAS to use multiple imaging sensors, simultaneously. In particular, thermal sensors combined with panchromatic imagery are incredibly beneficial for solar power inspections. The thermal images allow inspectors to identify specific deficiencies in the panels, and panchromatic imagery allows inspectors to identify physical damage or deterioration. Rather than having to manually check each panel one-by-one, an sUAS can capture the thermal signatures of 95,000 ft^2 of solar panels in roughly 45 min (Leggett 2019).

As a result of these real (and actualized) benefits, sUAS companies are designing vehicles and components specifically for infrastructure inspections. For example, PrecisionHawk (https://www.precisionhawk.com/) developed specific software and UAS workflows for solar, wind, and the oil and gas industry. Each of the developed packages is unique. For instance, PrecisionHawk includes thermal sensors for the solar monitoring package, electrical utility workflows include distribution lifecycles specific to electrical infrastructure, and thermal monitoring and methane detection come standard for the oil and gas package. DJI, the world leader in UAS manufacturing, recently released the Matrice 200 series sUAS for infrastructure inspections. It is designed to be both robust and rugged, coming standard with custom Zenmuse Z30 camera that boasts a 30× optical zoom for infrastructure integrity monitoring.

Other companies like ULC robotics out of New York (https://ulcrobotics.com), ABJ Drones (https://abjdrones.com/), and Intertek (http://www.intertek.com) all use third party platforms like PrecisionHawk or DJI, but the companies themselves focus on inspection services for the energy sectors. For example, consider RMUS, an industrial sUAS company specializing in the development of vehicles for specific hydrocarbon related industrial applications. They use the DJI Matrice M600 platform affixed with a specialized infrared camera capable of detecting a large range of hydrocarbon gases. Their complete system includes the hardware, software, and training and costs in the ball park of $85,000 USD (https://tinyurl.com/y35o5d7w). If nothing else, this price-point for an sUAS service provides readers with a good idea of how important drones are to the energy industry. In fact, the industry has grown so much that there is a dedicated Energy-Drone Robotics conference which was held in Houston in 2019 and "dedicated to launching, growing and scaling enterprise UAV/Robotics operations in energy companies."

The advantages of using sUAS for infrastructure monitoring, energy or otherwise, make a strong case for industry to further develop sUAS capabilities. At this juncture, however, one might argue that the most important advance in sUAS platforms has been the miniaturization of light detection and ranging sensors. The applications for LiDAR are numerous. Readers are probably most familiar with airborne LiDAR captured for large areas using specialized airplanes. However, the ability to fly LiDAR *on-demand, using an sUAS,* provides numerous benefits. First, the spatial accuracy of sUAS-mounted LiDAR is incredible. Consider, for example, the survey grade RIEGL VUX-1UAC sensor for UAS. The spatial accuracy is reported to be 10 mm (!), with a scan rate of 200/s and a point density of 40/m^2. To put this in perspective, larger airborne point densities collected by LiDAR sensors mounted to planes can be anywhere from 3 to 150 points per m^2 (Heidemann 2018; Wang et al. 2017; Zhu and Hyyppä 2014). Unfortunately, survey-grade LiDAR sensors for UAS sensors are expensive. The RIEGL VUX-1UAC retails for about $340,000 USD. A more "affordable" option, retailing for about $100 k, is the Geo-MMS Navigator VLP-32C designed by Geodetics Inc., which has a reported spatial accuracy of about 3 cm. As far as LiDAR resolution is concerned, 3 cm is excellent for typical applications and is much higher than most commercially or publically available data sets.

Another advantage of sUAS is the temporal consistency of data acquisition. Because sUAS operation only requires a minimum of two people (RPIC and spotter), a company or group with access to a LiDAR-equipped sUAS can capture data as frequently as desired. Even if a dedicated LiDAR sensor is not equipped on the sUAS, point clouds can still be generated by the on-board camera equipment and post-processing software. For example, imagery collected by the Ebee and Wingtra can be post-processed to create three-dimensional point clouds. These point clouds help to create digital elevation models (DEMs) and digital surface models (DSMs) and can be exported in the standard LAS (LiDAR data) file format (Fig. 8.1). Of course, the spatial accuracy will not match a LiDAR system, but sUAS point clouds are useful for a range of different applications (Rakha and Gorodetsky 2018), from capturing surface elevation, heights, slope and aspect, as well as calculating surface areas and volumes. When combined with orthophotos and imagery from other sensors (e.g., thermal, multispectral), a great deal of information is gained about the urban form.

Fig. 8.1 Point cloud generated in Pix4D using sUAS-derived imagery

8.3 sUAS Energy Infrastructure Monitoring and Assessment Research

For reasons unknown to the authors of this book, academic research pertaining to the use of sUAS for energy infrastructure monitoring is relatively sparse. The bulk of the research that does exist is found in computer science and engineering, primarily focusing on the development and use of sensors for power line inspections (Jalil et al. 2019) or on multi-vehicle coordination for inspections (Shi et al. 2018). Additional areas of research also include automated sUAS development for more efficient building inspection (Rakha and Gorodetsky 2018) and their use for inspections on energy structures.

As alluded to in the previous section, one of the advantages of using an sUAS for energy infrastructure inspection is the ability to identify problematic components or degradation without placing employees in dangerous situations. We also alluded to the financial advantages of replacing manual inspections with sUAS technology. One area where the financial and safety benefits of using sUAS are being realized is building energy audits. These types of audits are increasingly important in cities that deal with harsh seasonal climates and/or rely upon in-building heating and air conditioning. In a study conducted in Boston, Massachusetts, researchers found that upwards of 40% of energy lost in a residential area occurs through cracks and ducts that are not properly sealed (Shao 2011). In order to decrease energy loss from buildings, inspectors are tasked with the arduous process of manually testing all ducts, doors, and thermal bridges individually. This is both time consuming and expensive. Given this bottleneck, scholars are focusing on ways to develop sUAS technology to help automate this process.

For example, Rakha and Gorodetsky (2018) developed and tested a framework for using sUAS in automated building energy audits. The framework is based on field trials performed with a standard DJI multirotor UAV, providing a comprehensive account of flight patterns, image processing, and accuracy assessments. The authors also compare and contrast three different post-processing software packages and report on the usefulness of each for 3D structure modeling, as well as building energy loss analysis.

Another related area of sUAS infrastructure inspection research focuses on photovoltaic modeling, monitoring, and assessment. As mentioned previously, photovoltaic inspections for solar energy production have massive financial benefits (Leggett 2019). Thus, it is no surprise that researchers are developing different sUAS techniques for physically monitoring these energy systems (Gallardo-Saavedra et al. 2018). Some scholars are even exploring the possibility of retrofitting solar panel cleaning equipment to multirotor systems for automated cleaning (Al-Housani et al. 2019).

One of the most useful aspects of sUAS-derived data is their potential for transformation and use in 3D models. Again, this capability provides researchers and practitioners with accurate measurements for infrastructure integrity assessments, modeling, and design. While these point clouds are not strictly equivalent to

LiDAR-generated point clouds, they offer many of the same advantages. For example, each point in an sUAS point cloud contains an elevation, a geolocation, and an RGB color code. What is most exciting about these advances is that sUAS-derived data can now be used for most of the research and applications developed for traditional LiDAR data. While the quality and spatial accuracy of sUAS point clouds are variable, when sUAS data are combined with LiDAR, the monitoring and assessment of energy-related infrastructure can be improved.

8.3.1 LiDAR Applications for sUAS

A fair amount of the LiDAR-based infrastructure research deals with feature classification—from buildings and streets, to trees and utility poles. Not surprisingly, one of the more difficult features to extract from LiDAR is power lines because of their low surface area. However, several methods for extracting the spatial structure of power lines have been proposed. Some of these include rasterizing the LiDAR point clouds with classification based on height and point density (Zhu and Hyyppä 2014), as well as feature extraction using different neighborhood search scales for optimal identification (Wang et al. 2017). There is work directly related to sUAS in this domain. For example, Ippolito et al. (2016) present a voxel-based method for detecting power lines from on-board sUAS LiDAR systems. Interestingly, the goal was not to identify power lines for post classification. Rather, the research focused on the automatic detection of power lines for in-flight collision avoidance—an important consideration.

Applications for LiDAR extend far beyond the detection of power lines. As mentioned previously, many cities in the United States and elsewhere are developing strategies to become more sustainable, from cutting fossil fuel consumption to developing more efficient rubbish collection methods (Bibri and Krogstie 2017). Strategy development also includes estimating the solar energy capacities of cities. This process involves large investments in the installation of solar panels on municipal buildings, but also for commercial and residential buildings. Increases in solar capacity will not only provide power to the building on which it is installed but also contributes surplus electricity into the electrical grid (Nally 2017). An important consideration for assessing the feasibility of expanded solar capacity in a city is the evaluation of the urban landscape for its solar potential. Land is expensive (Florida 2017), so the idea of developing dedicated solar fields is not realistic for many cities. This means that the most logical place for solar installations are existing structures. LiDAR, especially when combined with orthoimagery, is one way to improve solar energy audits prior to any investments being made.

There are many *potential* applications of sUAS imagery for solar energy audits. For example, researchers have used remotely sensed land-cover and orthophotos to calculate the total usable surface area for solar panels on roof tops in a study region in eastern Ontario (Wiginton et al. 2010). In other work, researchers focused on calculating the total solar radiation of roof tops based on LiDAR-derived surface features (Kodysh et al. 2013), while others estimated the solar energy potential using building footprints, DEMs, airborne radar, and/or structural information (Hofierka and Kaňuk 2009; Li et al. 2015; Martín et al. 2015). When evaluating solar energy potential, there are three key elements that impact the amount of solar radiation that can be captured by roof top projects: (1) aspect, (2) slope, and (3) orientation. These elements are explored in detail by Lukač et al. (2013) who use LiDAR data to calculate total irradiance for a 1 km^2 urban area of Maribor, Slovenia. The results indicate significant variation in solar irradiance across the study area which the authors translated into potential solar energy generation.

What many of these articles either explicitly or implicitly communicate is the importance of high-resolution data describing the urban form. Airborne LiDAR offers many advantages, but it is costly to obtain and has limited temporal resolution. More importantly, recent work suggests that LiDAR data with a horizontal spacing of 1 m is too coarse to capture the necessary detail of urban features (Kodysh et al. 2013). In fact, Jia et al. (2013) suggest that LiDAR point cloud densities of 0.6 points/m^2 is the lower bound for generating the DEMs required for urban structure analysis. Given these known limitations of LiDAR, a major advantage of using sUAS for solar audits is their ability to capture data at extremely high resolutions, at reduced costs, when compared to city-wide airborne LiDAR data. For example, one of our most recent flights with the WingtraOne sUAS yielded a point cloud density of 1088.4 points/m^3 and a ground sampling distance of 1.36 cm. When combined with the digital orthophotos collected during the flight, all the ingredients for a comprehensive and high-resolution solar audit emerge. More importantly, since both the point cloud and photos are generated by the same sensor, they will exactly match—reducing horizontal and vertical errors that often propagate through the data when blending different data from different sensors.

To demonstrate the advantages of sUAS data for urban energy audits, we use data from a recent sUAS mission to estimate the solar potential for a residential neighborhood. There are many methods, frameworks, and workflows that can be used to conduct a solar audit, but in general, they all take a similar approach. In the following application, we are not setting out to reinvent the solar audit wheel. Rather, we seek to illustrate how sUAS-derived data can be used to perform a solar audit with high efficiency in a commercial geographic information system.

8.4 Urban Solar Audit Using an sUAS

Once again, the city of Phoenix, Arizona serves as our laboratory for this analysis. On average, Phoenix receives nearly 300 days of full sun, per year. Not surprisingly, Phoenix ranks sixth among all cities in the United States for installed solar capacity (Bradford et al. 2019). Our study focuses on the Mercury Mine neighborhood in north Phoenix. Aside from a local school and park, the neighborhood consists entirely of low-density, single-family homes (Fig. 8.2). Again, the advantage of using an sUAS for conducting an urban solar audit is the level of detail provided by

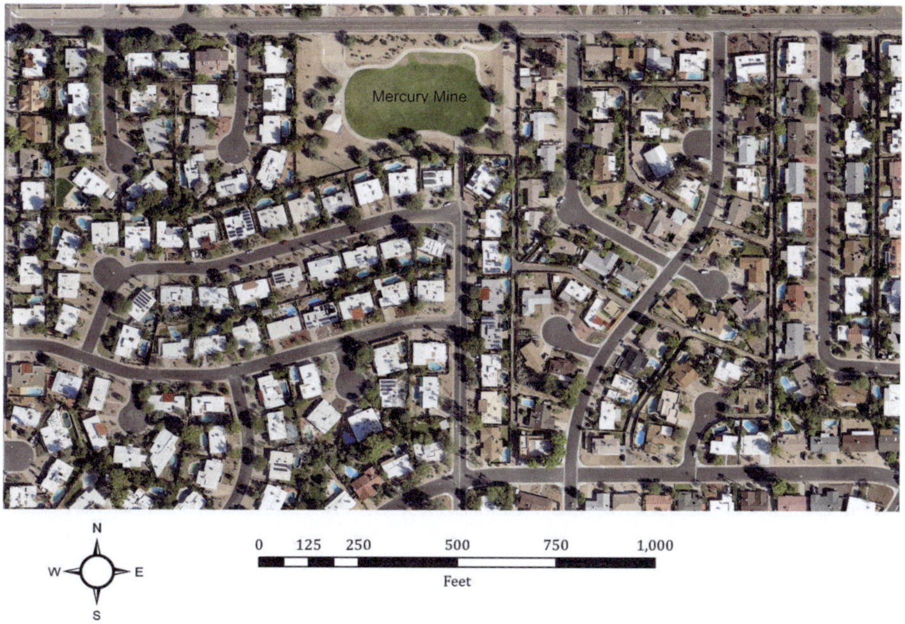

Fig. 8.2 Mercury Mine area of Phoenix, Arizona

the imagery. For the purposes of this application, we focus on determining the solar radiation of each roof top and the total potential surface area for energy production. Solar radiation is defined as a measure of the amount of solar energy received per unit area, often reported as watt hours (WH). Solar radiation may also be referred to as solar insolation or solar irradiance.[1]

Solar radiation depends on a number of factors including cloud cover, latitude, solar declination angle, zenith angle, and hour angle (Reda and Andreas 2004). Additionally, the three key elements of slope, aspect, and orientation must be accounted for. When taken together, these factors provide a path for determining the solar potential for roof tops in the study area. Once determined, one can estimate the total solar energy potential for all structures in a region.

[1] Of course, there are some technical differences in these terms (Boxwell 2012). Simply put, irradiance is to power, as insolation is to energy. Specifically, solar irradiance is the instantaneous measurement of solar power over an area. The typical measure is watts per square meter. Solar insolation is a measurement of the cumulative energy measured over an area for a defined period of time (e.g., weekly, monthly).

8.4.1 Mission Details

We flew the Mercury Mine mission in October, early in the morning using the WingtraOne sUAS, which is equipped with the Sony RX1RII 42-megapixel camera. The WingtraOne is a vertical take-off and landing (VTOL) platform that weighs approximately 10 pounds with the Sony payload. It has a wingspan of 125 cm and a cruising speed of nearly 36 mph. It can achieve GSD of 0.7 cm/px, but in practice, it routinely achieves between a GSD between 1 and 2 cm/px. When using post processing kinematics (PPK), absolute accuracies of 1 cm horizontal and 2 cm vertical can be achieved. Without PPK, the absolute accuracy is between 3 and 5 m.

The Mercury Mine mission covered 37.19 ha (~92 acres) and required 34 min of flight time. There were a total of 382 geotagged images which had a final ground sampling distance of 1.91 cm/pixel. We used Pix4D to post-process the imagery using a desktop computer with an Intel Core i7 processor at 2.8 GHz, 32 GB of RAM. The orthomosaic and digital surface model took 2 h to create. The final point cloud contained 58,918,152 individual points which translates into an average of 227.4 points/m^3.

As a brief aside, Pix4D, as well as other post-processing software, can classify points based on elevation, color, neighboring points, etc. This process yields an enhanced point cloud that eases one's ability to identify features across the study area, including buildings and vegetation. However, it is also important to consider the computational load that this requires. When the classification of points is left to Pix4D or other software, it takes time. Of course, the total processing time will vary based on the study area and the density of the points. Dedicated point cloud classification software, such as VRMesh (https://www.vrmesh.com/), is likely more efficient. The point cloud classification features in ArcGIS, housed within the 3D analyst toolbox, are another option, but our experience with these tools is not a positive one. If readers are software limited (i.e., ArcGIS only), we recommend starting by classifying ground points, noise, and then the buildings.

8.4.2 Data Preparation

As previously mentioned, several data layers are required for this analysis. First and foremost is a layer to represent the height of all features across the study area. Fortunately, in addition to stitching the natural color images together, Pix4D also generates a digital elevation model (DEM) from the associated point cloud. Again, one of the major benefits of using an sUAS for this type of analysis is the very fine spatial scale of the data collected. For example, the DEM displayed in Fig. 8.3 reflects a spatial resolution of 1.9 cm/pixel.

The second set of layers required for the solar audit are slope and aspect. Both of these are derived from the DEM using the spatial analyst toolbox in ArcGIS. For each pixel in the DEM, the slope of the surface is calculated in units of degree.

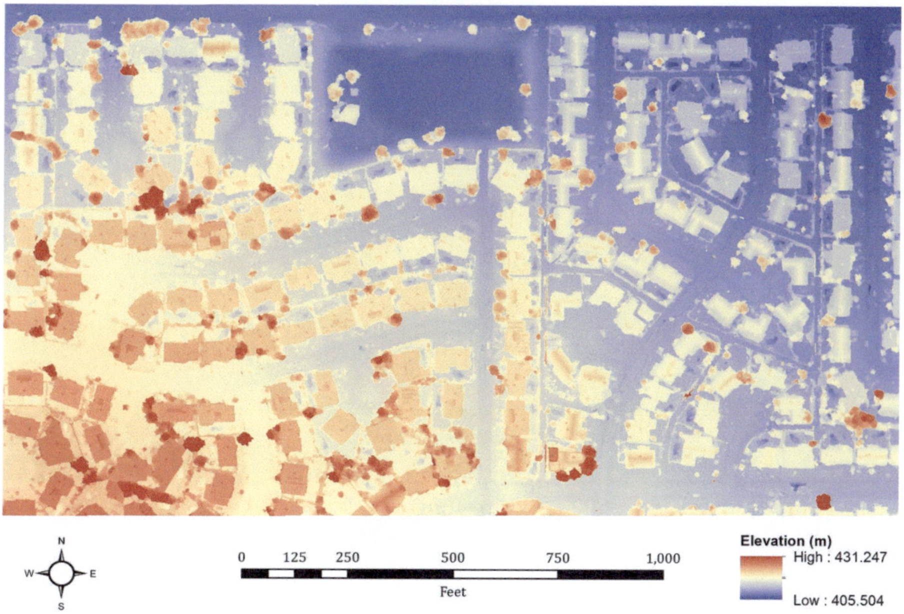

Fig. 8.3 Digital elevation model of Mercury Mine

Slope is simply the degree of incline of the surface. Aspect indicates the directional orientation of the surfaces in the study area. It is also recorded in degrees. However, because the two layers were created at such a fine spatial scale, a smoothing step is required to decrease the amount of noise on the surface. There are several options available for this. The first is to gently decrease the spatial resolution of the layer, being careful to not make it too coarse—obfuscating important surface information. The second is to use a smoothing filter (Gallant 2011). For the purposes of this analysis, the former worked best. The layers were resampled to 0.5 m^2 which was more than adequate to preserve the structural nuances. It also drastically reduced the computational burden during the final solar radiation calculations (Figs. 8.4 and 8.5).

At this juncture in the analysis, it is good practice to delineate which areas will be evaluated for solar potential. For example, if one was interested in placing solar panels in backyards, parks, open spaces, or any areas other than roof tops, life (and computing effort) becomes much easier and more efficient if one masks those areas prior to calculations for solar radiation. Since this analysis is focused on roof tops, we use a classified point cloud to identify and then generate building footprints. Ultimately, we use simple polygon shapefiles (black outlines) for clipping the slope and aspect layers so that only the structures were considered in the solar calculations. However, we leave the DEM intact for one critical reason—the final solar calculations need to account for the heights of all features in the data set. This is important because buildings, vegetation, or other features that shade roof tops need to be evaluated by the model to determine variations in solar radiation on structures.

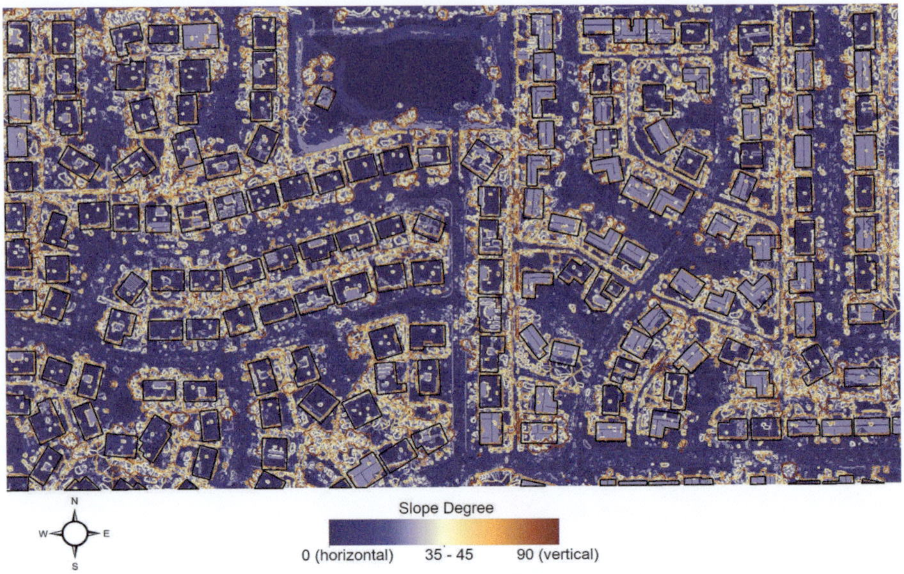

Fig. 8.4 Slope surface for Mercury Mine

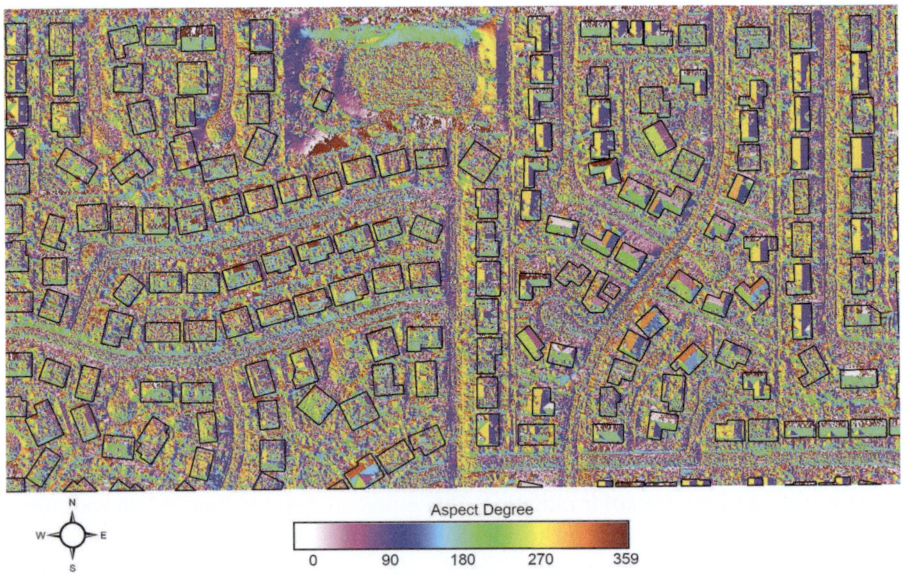

Fig. 8.5 Aspect surface for Mercury Mine

However, for larger study areas, or if computational effort is a concern, it is also possible to mask the DEM. If this option is pursued, we recommend buffering each building footprint by at least 15–30 feet to account for features surrounding the structures of interest (e.g., vegetation).

Our final step for preparing data for the solar audit, not surprisingly, is structured to minimize computational effort. Specifically, we can specify how the slope and aspect information are injected into the algorithm. This is accomplished by creating a point shapefile that jointly includes slope and aspect values of the rooftops across the study area. The unmasked DEM is still used as the input elevation surface. When combined with the point shapefile, we can now determine the final solar radiation estimates.

When considering the process of calculating solar potential, it is important to note that there is a large literature concerning methodological best practices. For example, there are continual efforts to improve methods for determining solar irradiance through better accounts of solar incident angles and changes in sun exposure across complex terrain (Šúri and Hofierka 2004). Other work focuses on the nuances of calculating solar potential in urban environments (see Freitas et al. (2015) for a recent review). That said, one of the most accessible and widely used methods of calculating surface radiation is the Solar Radiation Toolbox within ArcGIS. The recent work by Suomalainen et al. (2017) provides a great starting point for readers new to this research domain.

8.4.3 Analysis

Solar radiation can be calculated for single days, multiple days, or entire years (monthly intervals). The analysis presented here calculates solar radiation for a single, 9-h day (of sunlight) in December. We use a 1-h time interval for the solar estimates, meaning that the final shapefile (remember we are processing points and therefore the final output will also be points) will have nine attributes corresponding to the solar measurements for each hour of the day.[2] Once the final point-based output is generated, we convert them back to raster format for visualization and analysis. The solar potential raster is created at the same resolution as the original slope and aspect surfaces (0.5 m^2 in this case).

Since the data for this analysis is already in a projected coordinate system, the software automatically determines the latitude for solar declination and solar position. Within the ArcGIS tool, solar calculations are based on a viewshed algorithm developed by Rich et al. (1994). As part of the analysis, three components (rasters) are constructed: (1) a viewshed, (2) sky map, and (3) sun map. All three raster layers are then used to build an upward-looking hemispherical representation of the sky

[2] Readers should note that there is no point in running these calculations for more hours in a day than there is sun.

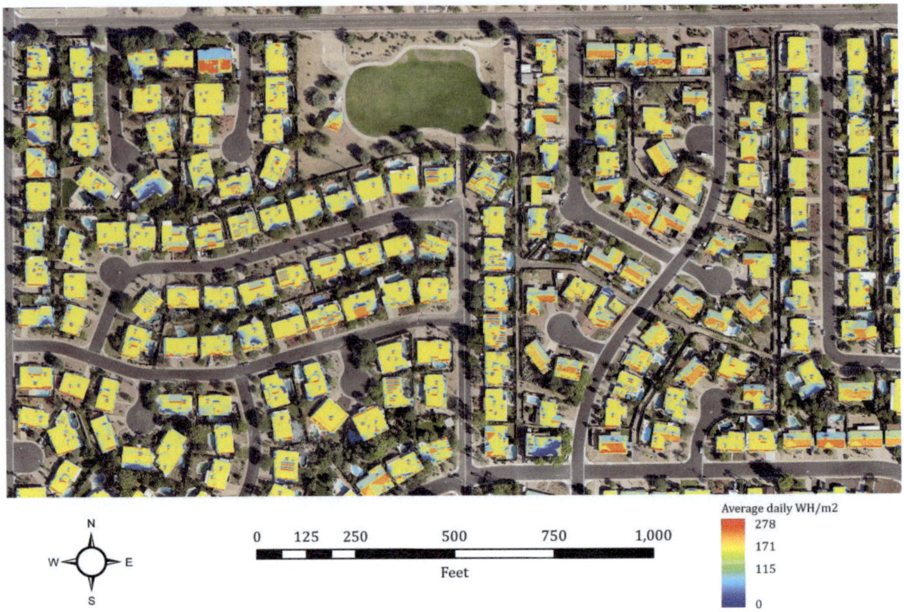

Fig. 8.6 Daily average global solar radiation estimates in December for Mercury Mine

above the study area. The default value within the tool, 200 m × 200 m, was used with good results. Increasing the resolution of the viewshed raster is possible and will also result in higher overall accuracy, but will drastically increase processing time.

Figure 8.6 illustrates the daily average *global* solar radiation estimates in December for all roof tops across the study area. Global radiation is the sum of diffuse radiation (light that is scattered) and direct radiation for all sun map and sky map sectors. The final estimates are reported as watt hours per square meter (WH/m^2). The resulting values reflect the number of watt hours per square meter that are produced over the specified temporal window. Astute readers will likely notice some interesting features in the two images that contribute to higher (red colors) or lower (blue colors) solar radiation. To this extent, there are several points worth discussing in more detail.

As you would expect, the orientation and shape of the rooftop play a significant role in solar radiation. Most structures in the western half of the study area have flat roofs, while many of the structures on the eastern half of the study area have pitched roofs. As a result, the homes on the east side have less solar power potential, especially in the winter months when the sun is lower in the sky (Fig. 8.7). Also noticeable is the effect of vegetation on solar potential. For example, [*Item 1*] in Fig. 8.7 displays a house with a large dark blue (low radiation) patch caused by a cluster of large trees surrounding the south side of the structure. With this information in hand, a planner could easily modify the urban landscape to maximize roof exposure to incoming solar energy. Of course, in Phoenix, these trees also provide substantial

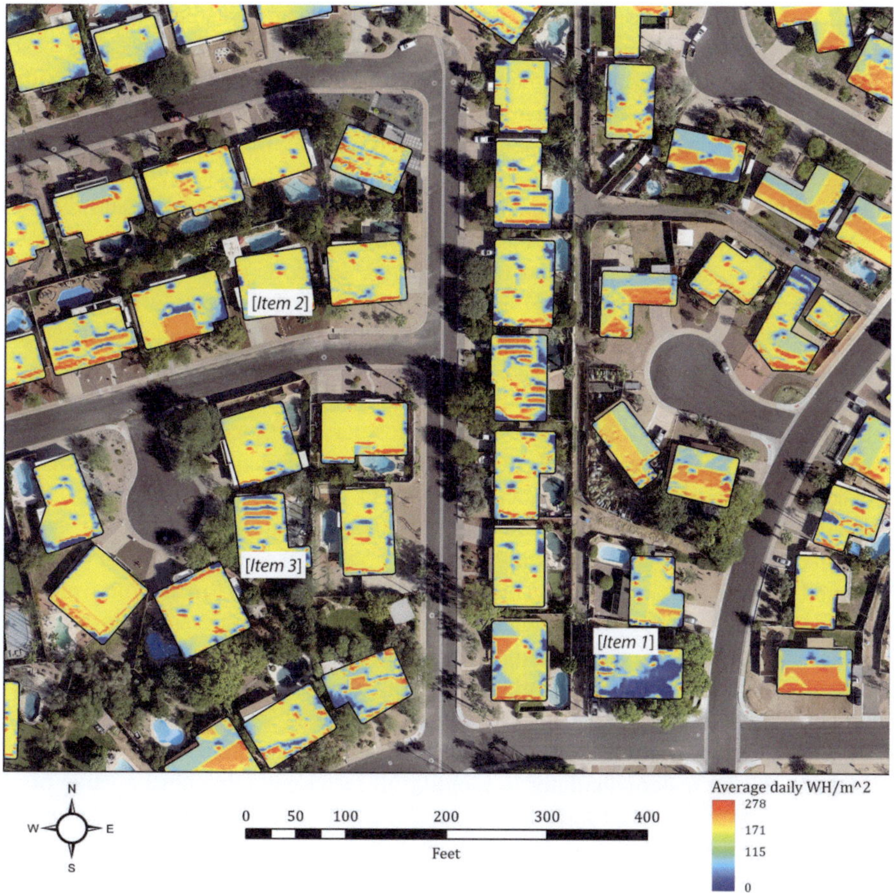

Fig. 8.7 A closer look at daily average global solar radiation estimates for Mercury Mine

shading (and cooling) in the summer, so there is a trade-off to be considered by the homeowner. Nevertheless, by planning for roof pitch and optimally spacing vegetation, the solar radiation on structure tops can be maximized. Furthermore, planners could also use this information to identify optimal locations for solar panels, given the structure and layout of the urban landscape.

The calculated solar radiation also speaks to the level of detail provided by the sUAS imagery. Across all the rooftops, there is noticeable variation in the amount of solar radiation. Readers may notice that most rooftops display small areas of lower solar potential, backing small areas of high solar potential. These are rooftop air conditioning units (*Item 2*). And yes, the temperatures are hot enough in Phoenix that most houses require multiple air conditioning units on their roof. Also interesting are the candy-striped (high/low) patterns found on some rooftops in Fig. 8.7. *Item 3* highlights one example, corresponding to a house that already has solar panels installed. The lines of high solar radiation are the angled solar panels and the lines of low solar radiation immediately behind them are the shadows they cast. Given this level of detail, urban planners, energy specialists, or sustainability officers can not only detect whether a house already has solar panels, but also determine whether the solar panels are optimally situated to capture the most amount of solar radiation in a day, month, or year. In addition, the ability to conduct these types of solar audits (annually or semiannually) provide important time-series data regarding urban sustainability goals that cannot be matched by other data sources. Without the fine scale resolution provided by the sUAS, some of these nuances may be obfuscated or completely missed. For example, the lack of high-resolution elevation data can alter the total estimate of shaded roof tops which will ultimately impact the accuracy of solar radiation estimates.

The total surface area available for solar panel installations can easily be calculated from the information we now have. The solar radiation values (displayed in Fig. 8.7) identify surfaces with high solar potential, and this is followed by a simple surface area calculation for each location. For this analysis, any surface that has an average daily solar radiation value above 750 WH/m^2 is considered viable. It should be noted, however, that solar panel efficiency with respect to input-output will change based on the time of year, the type of solar panel, and the cloud cover. The value of 750 selected for this application is arbitrary and only serves to illustrate the process. We also use the winter global radiation raster for this application. Remember, we conducted this flight in December, when the sun angle and associated solar radiation values are lower. One could conduct a similar audit during the spring, summer, and fall to help optimize solar panel placement for neighborhood structures. Using raster calculator to identify all raster cells that have a value of at least 750 WH/m^2, the resulting surface area for potential solar panel installations is 0.051 km^2 (Fig. 8.8).

It is now possible to estimate the amount of power this neighborhood could generate given the potential solar surface area of each structure and the calculated daily average WH/m^2. Consider the generalized equation for calculating the energy produced by a solar panel (Šúri and Hofierka 2004):

$$E = A * r * H * \mathrm{PR} \tag{8.1}$$

A is the total area of the roof tops, r is the solar panel yield (in percent) which is calculated by the ratio of electrical power rating for the pane divided by the area of the panel (we will assume 20% since that is the mid-point of a typical operating yield[3]), H is the average daily solar radiation on the panels, and PR is the performance ratio. PR reflects the context specific quality features of the solar panel system. For the purposes of this analysis, we borrow the list of loss factors and their associated percentages from the Photovoltaic Software website[4] that include (Table 8.1):

---

[3] https://news.energysage.com/what-are-the-most-efficient-solar-panels-on-the-market/

[4] https://photovoltaic-software.com/principle-ressources/how-calculate-solar-energy-power-pv-systems

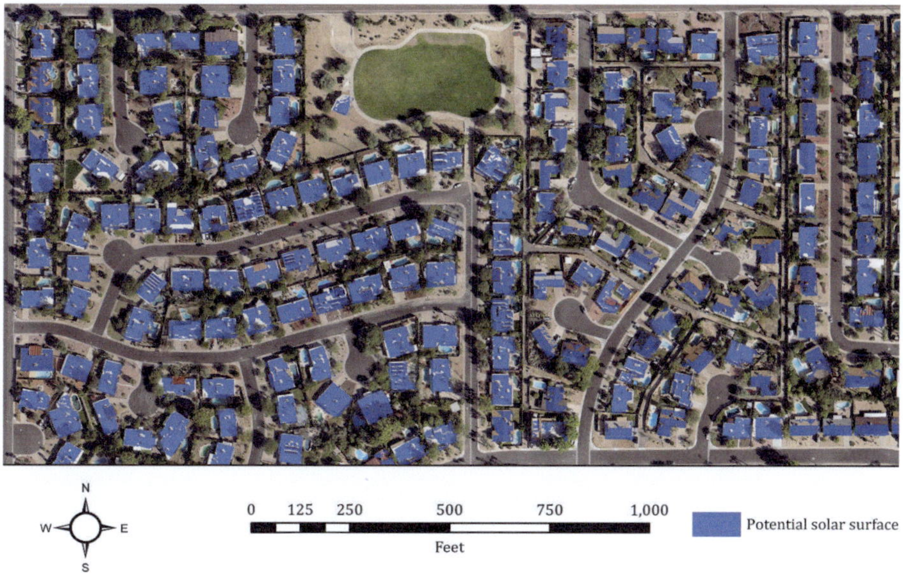

Fig. 8.8 Potential solar surfaces for Mercury Mine

Table 8.1 Several of the most important variables that contribute to solar panel efficiency loss. These are averages and could vary from system to system

Category and potential loss value	Efficiency values used (%)
Inverter losses (IL) (6–15%)	9
Temperature losses (TL) (5–15%)	0
DC cables losses (DC) (1–3%)	2
AC cables losses (AC) (1–3%)	2
Shadings (SH) 0–40% (depends on site)	0
Losses weak irradiation (WIR) 3–7%	3
Losses due to dust, snow, etc. (DUST) (2%)	2
Performance ratio	**0.84**

An average value is used for capturing inverter losses, DC cable losses, AC cable losses, and losses from weak irradiation. For this analysis, temperature losses are 0% since the optimal operating temperature of most solar panels is 77 °F and the average temperature in Phoenix in December is 65 °F. We previously identified all the areas with optimal sun exposure so shading loss is negligible for these surfaces. Given the desert landscape, Phoenix is quite dusty, so we apply the max loss of 2%. The performance ratio is then calculated using the following formula:

$$PR = (1-IL)*(1-TL)*(1-DC)*(1-AC)*(1-SH)*(1-WIR)*(1-DST) \quad (8.2)$$

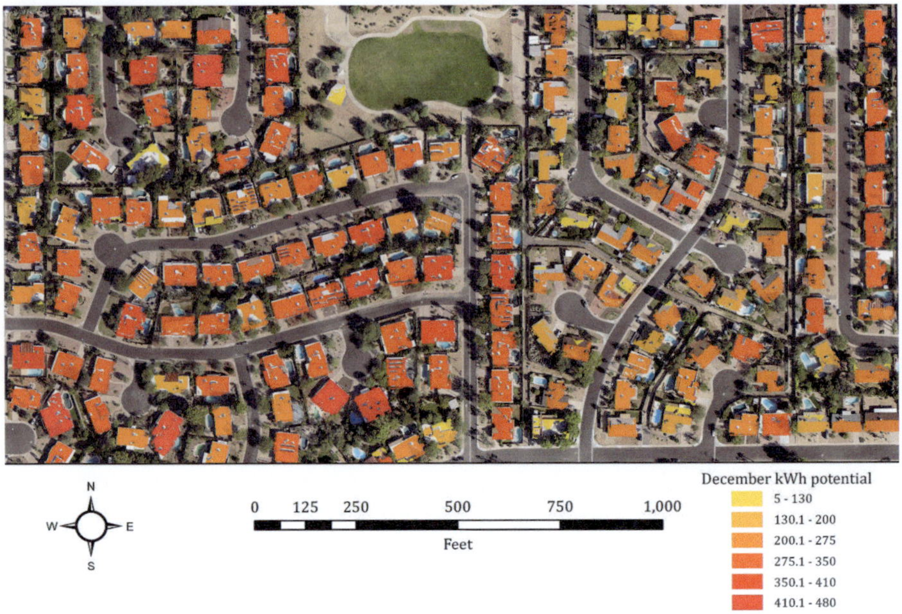

Fig. 8.9 The total December solar potential for each rooftop

After calculating the PR and entering it into (Eq. 8.1), the resulting units represent average WH/m^2 for a single day in December. We estimate the total kWh for the entire month of December by multiplying the values of *E* by 0.001 (the conversion to kilowatts) and then by 31 (representing each day of the month). Figure 8.9 formally illustrates the spatial configuration of the results. If we fit every eligible surface in the neighborhood with panels, after carefully filtering out extraneous rooftop areas that are under ~20 ft^2 (roughly the size of one solar panel, usually a shed or other small structure), the total amount of energy produced (on average) in December is ~47,507.40 kWh.

To readers unfamiliar with kilowatt or kilowatt hour measurements, the 47,507 kWh may seem like a lot of energy. But, let us put this measurement in perspective. First, it is important to understand the difference between a kilowatt and a kilowatt hour. A kilowatt is simply 1000 W. For example, if one owns a 1000-W microwave, it requires 1000 W (1 kW) of power to operate. If that same microwave was operating for an hour straight, it would consume 1 kWh of energy. The United States Energy Information Administration (EIA) reports that the average household in the United States consumes about 900 kWh of energy per month (EIA 2018). The house with the largest solar energy potential in Mercury Mine for December (478.25 kWh) could only cover half of the energy used by an average US household. In short, readers should not let the aggregate statistic of 47,507 kWh distract them from the bottom line. The neighborhood could generate a substantial amount of energy, but nowhere near enough to be self-sufficient for December. Again, this is not surprising given the sun angle.

8.5 Limitations and Implications of Using sUAS for Energy Infrastructure Assessment

sUAS are beneficial for energy infrastructure monitoring and assessment in several ways. First, they can be used to safely and efficiently inspect faulty equipment. Whether this is a large solar panel array, or a difficult-to-reach infrastructure component like a wind turbine or long stretches of pipeline in rough terrain, sUAS add both speed and accuracy to energy infrastructure evaluations. sUAS-derived data also help create insightful geovisualizations of infrastructure systems, including 3D models that allow planners, practitioners, and researchers, to better evaluate the infrastructure in near-real time. Third, the 3D data combined with orthoimagery provides the geospatial intelligence required for estimating solar potential for complex urban landscapes. Throughout this chapter, we demonstrated several of the advantages bestowed by sUAS for energy infrastructure assessments, with a focus on solar energy potential in a residential neighborhood. The analytical framework detailed here is easily extendable to other locations, helping planners determine the best possible allocation of solar panels for energy development. As detailed above, shading, slope, aspect, and rooftop orientation play important roles in solar energy potential and panel efficiency. Analyses performed at suboptimal spatial resolutions can lead to inefficient allocations of solar equipment and low-efficiency solar installations.

It is important to note that the analytical framework detailed in this chapter can also be applied to passive cooling analysis. In many of the warmest communities here in the United States, such as Phoenix, Austin, and Atlanta, summer months can produce temperatures that exceed 100 °F for multiple, consecutive weeks. Air conditioners are a necessity. However, air conditioners require significant amounts of energy for operation, which can strain local power grids when the temperatures soar. One approach for reducing energy needs and consumption in an urban landscape is to place vegetation around structures to shade them during the hotter parts of the day. In places like Phoenix, this might include planting palo verde trees, as an example, on the south side of one's house. As the trees grow, the shade provided by their foliage reduces the total amount of solar radiation received by the house, decreasing energy requirements for cooling. The data generated from sUAS can guide these types of analyses. At the same time, this shade may also reduce the efficiency of installed solar panels. This is a trade-off that needs to be evaluated in future research.

Finally, when considering the different approaches for conducting energy audits, the benefits and costs associated with both LiDAR and sUAS must be considered. The cost to obtain LiDAR data can be enormous. As cities continue to transform, adding (or razing) housing, commercial buildings and the like, the structural footprints of urban areas change constantly. Airborne LiDAR data collected in 2019 might be representative for 2020, but in the years that follow, it is unlikely to remain so. sUAS-derived data provide a bridge, helping to update geospatial intelligence portfolios with a blend of high-resolution orthophotos and 3D point clouds—all of which can help urban energy initiatives.

Although there are many benefits to sUASs for energy assessment, there are several limitations that readers should be aware of. First, although flight durations for sUAS are increasing, the geographic range of hyper local analyses remain limited. The analysis presented in this chapter covered roughly 100 acres, but could be extended to upwards of 200 or 250 acres with a single flight. However, extending the study area beyond 250 acres would require multiple missions. This takes time, coordination, and a lot of fresh batteries. Second, while we tend to stress the value and importance of super high-resolution data, processing these data takes time. Our analyses are conducted on high-end desktop computers. Readers with more modest computing capabilities might find the performance bottlenecks frustrating. While we offered some efficient workarounds in this chapter for handling large databases, including resampling to coarser resolutions, the data remain large and computing times long. We encourage readers to explore creative alternatives to these types of energy assessments using sUAS.

References

Al-Housani, M., Bicer, Y., & Koç, M. (2019). Experimental investigations on PV cleaning of large-scale solar power plants in desert climates: Comparison of cleaning techniques for drone retrofitting. *Energy Conversion and Management, 185*, 800–815.

Bibri, S. E., & Krogstie, J. (2017). Smart sustainable cities of the future: An extensive interdisciplinary literature review. *Sustainable Cities and Society, 31*, 183–212.

Boxwell, M. (2012). *Solar electricity handbook*. Warwickshire: Greenstream Publishing.

Bradford, A., Stankiewicz, T., Sundby, J., Fanshaw, B., & Sargent, R. (2019). *Shining cities 2019*. Washington, D.C.

Florida, R. (2017). *The staggering value of urban land*. City Lab. Retrieved July 8, 2019, from https://tinyurl.com/y2y7hesx

Freitas, S., Catita, C., Redweik, P., & Brito, M. C. (2015). Modelling solar potential in the urban environment: State-of-the-art review. *Renewable and Sustainable Energy Reviews, 41*, 915–931.

Gallant, J. (2011). Adaptive smoothing for noisy DEMs. *Geomorphometry, 2011*, 7–9.

Gallardo-Saavedra, S., Hernández-Callejo, L., & Duque-Perez, O. (2018). Technological review of the instrumentation used in aerial thermographic inspection of photovoltaic plants. *Renewable and Sustainable Energy Reviews, 93*, 566–579.

Ham, Y., Han, K. K., Lin, J. J., & Golparvar-Fard, M. (2016). Visual monitoring of civil infrastructure systems via camera-equipped Unmanned Aerial Vehicles (UAVs): A review of related works. *Visualization in Engineering, 4*(1), 1.

Heidemann, H. K. (2018). *Lidar base specification*. Washington, D.C.

Hoffmann, G. M., Huang, H., Waslander, S. L., & Tomlin, C. J. (2011). Precision flight control for a multi-vehicle quadrotor helicopter testbed. *Control Engineering Practice, 19*(9), 1023–1036.

Hofierka, J., & Kaňuk, J. (2009). Assessment of photovoltaic potential in urban areas using open-source solar radiation tools. *Renewable Energy, 34*(10), 2206–2214.

Ippolito, C., Krishnakumar, K., & Hening, S. (2016). Preliminary results of powerline reconstruction from airborne LiDAR for safe autonomous low-altitude urban operations of small UAS. In *IEEE Sensors* (pp. 1–3).

Jalil, B., Leone, G. R., Martinelli, M., Moroni, D., Pascali, M. A., & Berton, A. (2019). Fault detection in power equipment via an unmanned aerial system using multi modal data. *Sensors, 19*(13), 3014.

Jia, Y., Lan, T., Peng, T., Wu, H., Li, C., & Ni, G. (2013). Effects of point density on DEM accuracy of airborne LiDAR. In *International Geoscience and Remote Sensing Symposium-IGARSS* (pp. 493–496). Melbourne: IEEE.

Kanellakis, C., Fresk, E., Mansouri, S. S., Kominiak, D., & Nikolakopoulos, G. (2019). *Autonomous visual inspection of large-scale infrastructures using aerial robots*. arXiv preprint arXiv:1901.05510.

Kodysh, J. B., Omitaomu, O. A., Bhaduri, B. L., & Neish, B. S. (2013). Methodology for estimating solar potential on multiple building rooftops for photovoltaic systems. *Sustainable Cities and Society, 8*, 31–41.

Leggett, T. (2019). Using infrared drone technology to inspect solar panels and wind turbines. *Power Magazine*. Retrieved July 6, 2019, from https://www.powermag.com/using-infrared-drone-technology-to-inspect-solar-panels-and-wind-turbines/

Li, Z., Zhang, Z., & Davey, K. (2015). Estimating geographical PV potential using LiDAR data for buildings in downtown San Francisco. *Transactions in GIS, 19*(6), 930–963.

Lukač, N., Žlaus, D., Seme, S., Žalik, B., & Štumberger, G. (2013). Rating of roofs' surfaces regarding their solar potential and suitability for PV systems, based on LiDAR data. *Applied Energy, 102*, 803–812.

Martín, A. M., Domínguez, J., & Amador, J. (2015). Applying LIDAR datasets and GIS based model to evaluate solar potential over roofs: A review. *AIMS Energy, 3*(3), 326–343.

Mason, E. (2018). Drone-based thermal imaging and analytics for energy efficiency. *Phys.org*. Retrieved June 24, 2019, from https://tinyurl.com/y9a7abd7

Nally, B. (2017). Is solar power the future for smart cities? *Smart Cities Dive*. Retrieved July 8, 2019, from https://tinyurl.com/yxcl8j9w

Oung, R., & D'Andrea, R. (2011). The distributed flight Array. *Mechatronics, 21*(6), 908–917.

Rakha, T., & Gorodetsky, A. (2018). Review of Unmanned Aerial System (UAS) applications in the built environment: Towards automated building inspection procedures using drones. *Automation in Construction, 93*, 252–264.

Reda, I., & Andreas, A. (2004). Solar position algorithm for solar radiation applications. *Solar Energy, 76*(5), 577–589.

Redweik, P., Catita, C., & Brito, M. (2013). Solar energy potential on roofs and facades in an urban landscape. *Solar Energy, 97*, 332–341.

Rich, P., Dubayah, R. C., Hetrick, W., & Saving, S. (1994). Using viewshed models to calculate intercepted solar radiation: Applications in ecology. In *American Society for Photogrammetry and Remote Sensing Technical Papers* (pp. 524–529). American Society of Photogrammetry and Remote Sensing.

Shao, E. C. (2011). *Detecting sources of heat loss in residential buildings from infrared imaging*. Massachusetts: MIT Libraries.

Shi, W., Zhou, H., Li, J., Xu, W., Zhang, N., & Shen, X. (2018). Drone assisted vehicular networks: Architecture, challenges and opportunities. *IEEE Network, 32*(3), 130–137.

Suomalainen, K., Wang, V., & Sharp, B. (2017). Rooftop solar potential based on LiDAR data: Bottom-up assessment at neighbourhood level. *Renewable Energy, 111*, 463–475. Retrieved from. http://www.sciencedirect.com/science/article/pii/S0960148117303324.

Šúri, M., & Hofierka, J. (2004). A new GIS-based solar radiation model and its application to photovoltaic assessments. *Transactions in GIS, 8*(2), 175–190.

United States Energy Information Association [EIA]. (2018). How much electricity does an American home use? *FAQ*. Retrieved September 20, 2019, from https://www.eia.gov/tools/faqs/faq.php?id=97&t=3

Vian, J. L., Mansouri, A. R., & Saad, E. W. (2011, November 15). *System and method for inspection of structures and objects by swarm of remote unmanned vehicles*. US Patent Number US8060270B2. Retrieved from. https://patents.google.com/patent/US8060270

Wang, Y., Chen, Q., Liu, L., Zheng, D., Li, C., & Li, K. (2017). Supervised classification of power lines from airborne LiDAR data in urban areas. *Remote Sensing, 9*(8), 771.

Wiginton, L. K., Nguyen, H. T., & Pearce, J. M. (2010). Quantifying rooftop solar photovoltaic potential for regional renewable energy policy. *Computers, Environment and Urban Systems, 34*(4), 345–357.

Zhu, L., & Hyyppä, J. (2014). Fully-automated power line extraction from airborne laser scanning point clouds in forest areas. *Remote Sensing, 6*(11), 11267–11282.

Chapter 9
Detecting and Monitoring Informal Trails in an Urban Mountain Preserve Using Small Unmanned Aerial Systems

Abstract Informal trails represent a serious threat to open spaces, parks, wilderness areas, and protected natural areas in many different locations and contexts. From vegetation loss, the displacement of wildlife, and alteration of habitats to soil degradation and erosion, landscape fragmentation, and the spread of invasive species, significant and continual effort to detect, monitor, and interdict informal trails is required. The purpose of this chapter is to detail the ways in which small unmanned aerial systems can contribute to efforts for detecting the presence of informal trails. A case study of the Phoenix Mountains Preserves in Phoenix, Arizona helps to illustrate how basic spatial analytics combined with sUAS imagery can identify landscape fragmentation and potential interdiction points for minimizing the use of informal trails for enhancing landscape recovery.

Keywords Informal trails · Preserve · Soil · Wildland · Piestewa · ArcGIS · FRAGSTATS · Field inventory

9.1 Introduction

By definition, informal trails are visually discernable pathways that are not part of the formal trail system, not maintained and/or not recognized by authorities charged with managing recreational space (e.g., parks, wilderness areas, protected natural areas). There is no single reason motivating the creation of informal trails. In some instances, informal trails represent a more direct path (i.e., a shortcut) between access points and areas of interest. In other instances, the informal trails are created by animals (e.g., deer), pack animals, or all-terrain vehicles (Leung et al. 2011; Hammitt and Cole 1998, Hammitt et al. 2015; Kutiel 1999; Pickering and Norman 2017). Only later are the trails used by humans.

The problems with informal trails are multifaceted. First, the ecological impacts can be severe. Mechanical disturbances (e.g., stepping on/trampling) to highly sensitive vegetation, including vascular plants and lichen (Pickering and Growcock

T. H. Grubesic, J. R. Nelson, *UAVs and Urban Spatial Analysis*,
https://doi.org/10.1007/978-3-030-35865-5_9

2009; Pescott and Stewart 2014; Jägerbrand and Alatalo 2015) create a range of undesirable effects, including a reduction in cover and increasing opportunities for invasive species and/or weeds (Jägerbrand and Alatalo 2015). This is particularly true in alpine environments, where the growing season is short and recovery times are slow for native plants (Willard et al. 2007). Second, informal trails fragment larger, intact landscapes into smaller patches. In these scenarios, soil compaction can reduce water holding capacity, altering precipitation run-off patterns, increasing soil erosion and ultimately degrading local habitats. More importantly, the informal trails can function as barriers for the movement of small animals and insects, while invasive species often use the trails as conduits into sensitive ecological regions (Leung et al. 2011; Forman 1995; Forman and Alexander 1998). Third, informal trails can also increase probabilities of vandalism and looting from important cultural and environmental sites (Leung et al. 2011). For example, in the state of Oregon, thieves steal thousands of pounds of plants, moss, and fungi on an annual basis, selling it through illegal channels (Odegard 2014). Law enforcement officials recently seized 13,000 pounds of salal (a bushy, green-leafed plant that is used in floral arrangements and valued for its long shelf-life) from a warehouse in Philomath, Oregon. Destined for the east coast or Europe, the salal had a value of more than $15,000. Thieves access the salal patches via logging roads and informal trails located in protected wilderness areas throughout the state. In Arizona, the story is similar, but thieves target cacti instead (McGivney 2019).

One outstanding challenge for environmental managers is the process of identifying, assessing, and interdicting the development of informal trails in sensitive landscapes. As detailed by Marion et al. (2006), informal trail networks can be extensive, making the efficiency of field assessments paramount. Leung et al. (2011) provide a basic typology of approaches for monitoring informal trails: (1) *estimates*, (2) *field inventory*, and (3) *remote sensing*. Estimates are the least accurate and typically use surrogate measures for extrapolating informal trail systems. For example, this may consist of counting the number of paths originating from local campgrounds or campsites, or using line intercept methods where informal trail crossings are counted along the formal trail system (Boorman and Fuller 1977). Once viable, this type of approach is not particularly accurate, telling managers very little about the trail characteristics (e.g., length, width, density). Field inventory methods, which often include the mapping and complete census of informal trail networks using ground surveys, are popular and reliable, but time consuming and expensive. This approach often makes use of global positioning systems (GPS) or GPS-enabled smart phones for capturing the trails (Barros and Pickering 2017; Korpilo et al. 2017; Leung et al. 2011; Walden-Schreiner and Leung 2013). In turn, the GPS coordinates are integrated into a geographic information system (GIS) for the mapping and analysis of the trails detected. Lastly, it is possible to identify and delineate informal trails by using remotely sensed imagery. This can include digital orthophotography from an airborne platform, including satellites, planes, or unmanned aerial systems (UAS) (Barros and Pickering 2017; Martin et al. 2018; Westcott and Andrew 2015; Wimpey and Marion 2011). One can easily integrate these images into a GIS for analysis and evaluation of the informal trails.

The problem with using commercial imagery for the detection and evaluation of informal trails is threefold. First, the imagery can be expensive. For example, WorldView 2/3 panchromatic images from LandInfo (a satellite imagery vendor) cost \$14 per km^2, with a minimum order of 25 km^2 with a 2 km width. This equates to an order of \$700 USD (LandInfo 2018) for older, archived data. Costs increase dramatically if users want to task the satellite for collecting current data (\$48 per km^2 with a minimum order of 100 km^2). Second, the temporal signature of satellite data can be difficult to ascertain. In a recent work, Westcott and Andrew (2015) use Google Earth imagery (free), estimating that the images were collected between 2006 and 2012. This is a relatively large temporal spread (8 years), which can introduce significant uncertainty into the analysis. Third, it can be difficult to integrate imagery collected at multiple times, by multiple sensors, into a coherent mass for analysis.

Given the strengths and weaknesses associated with traditional approaches for identifying and monitoring informal trails, unmanned aerial systems represent an innovative, accurate, and highly efficient alternative for collecting crucial geospatial intelligence in protected and/or recreational areas. UAS can cover hundreds of acres in a few hours and provide imagery with sub-centimeter resolutions—drastically reducing the need for dedicated man-hours walking the length of informal trails with a GPS for mapping purposes. Further, because the resolution of UAS-derived imagery is so high, evaluation of trail geometries, densities, soil conditions, habitat disturbance, fragmentation, and the like can be evaluated with precision and accuracy.

The purpose of this chapter is to detail the ways in which unmanned aerial systems can contribute to efforts for detecting the presence of informal trails in protected and/or recreational areas. A case study of the Phoenix Mountains Preserves in Phoenix, Arizona helps to highlight the utility of sUAS deployment and imagery for identifying informal trails. In addition, we use a suite of basic spatial analytical approaches combined with sUAS imagery to identify potential interdiction points for minimizing the use of informal trails for enhancing landscape recovery.

9.2 Phoenix Mountains Preserves

The Sonoran Desert is one of the most ecologically diverse deserts on the planet. With a geographic expanse of 260,000 square miles, the desert consists of numerous eco-regions and is home to more than 100 reptile species, 60 different mammals, 350 bird species, and 2000 different plant species, including the Saguaro Cactus (Hawes 2005). The fifth largest (and fastest growing) city in the United States, Phoenix, Arizona (pop. 1,660,272), is in the heart of the Sonoran Desert (Gallen 2019). Between 2017 and 2018, Phoenix grew by approximately 70 people per day, adding a total of 25,288 residents (Gallen 2019). One of the reasons that Phoenix is such an attractive destination for people is the abundance of recreation opportunities in/around the metropolitan area. Specifically, Phoenix is home to 37,000 acres of

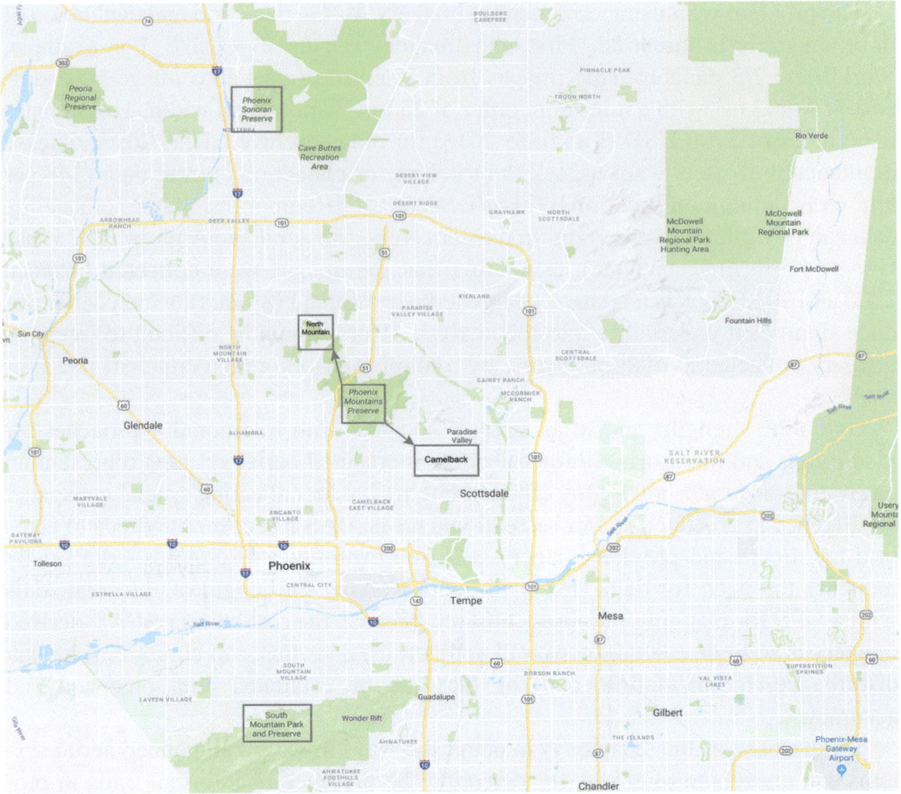

Fig. 9.1 Phoenix Mountains Preserves, Phoenix, Arizona

mountain preserves, consisting of North Mountain/Shaw Butte, Camelback Mountain, Piestewa Peak, and the Phoenix Sonoran Preserve (Fig. 9.1). The preservation process for these areas began in 1924 with the purchase of 16,283 acres consisting of South Mountain—an area considered to have "great recreational value" (Rothwell 2014). Through the years, additional land was added, expanding the preserves beyond South Mountain, but this process has proven controversial for a variety of reasons. Kiefer (1998) and Rothwell (2014) provide additional context regarding the constant, and relatively uneasy relationship between the preserve(s), rapid population growth and associated development pressures of Phoenix area.

The proximity of the mountain preserves to local population centers (e.g., Phoenix, Scottsdale, Mesa, Chandler) means that they are heavily trafficked. For example, about 750,000 people use the trails in South Mountain Park each year, with an additional 250,000 cars driving through the park. Estimates for Camelback Mountain, one of the most popular hikes in the entire state of Arizona, suggest that 450,000 visitors hike its two trails annually (Dale 2018).

There are multiple problems with this type of heavy traffic in preserve areas. For example, on May 26, 2016 at 1:30 am, an emergency rescue crew and a helicopter were dispatched to North Mountain Park to help a distressed hiker down the mountain. Later, the man revealed to rescuers that he arrived to the preserve on a city bus to hunt snakes.[1] The man did not bring any water with him, but he did have some beer (Wahl 2016). Although the combination of snake hunting and beer consumption is amazing, it is the sheer number of people making their way through the preserves daily that creates serious sustainability concerns for the sensitive ecosystems found throughout.

9.2.1 Sensitive Desert Soils

Damage to desert soils is one of the larger concerns for the Phoenix Mountains Preserves. McAuliffe (1994) details the sensitive nature of these soils, particularly the role of calcic horizons which are literally the "glue" for many desert soils. Calcic horizons are accumulations (in variable thicknesses) of calcium carbonate. In the Sonoran Desert, the horizons are found at varying depths below the soil surface. In wetter regions (e.g., Tucson), the calcium carbonate begins to accumulate at 10 in. or deeper. In more arid regions (e.g., Yuma), it may only be a centimeter or two below the soil. No matter the location, thicker calcic horizons are slow to form, often taking hundreds of thousands of years to collect. Mechanical disturbances to calcic horizons can create long-term damage to local ecosystems, including changes to water run-off regimes. McAuliffe (1994) also notes that the microphytic crusts found throughout the Sonoran Desert are crucial to long-term ecosystem stability. Microphytic crusts consist of cyanobacteria, algae, liverworts, lichens, and mosses that help retain water after rainfall, knit the soil together (helping soils to resist erosion), and fix nitrogen to help larger plants to grow. When these crusts are mechanically disturbed by human feet, pack animals, mountain bikes, or all-terrain vehicles, the recovery times may take several decades or perhaps a century or more (McAuliffe 1994). Not surprisingly, the loss of these crusts can have downstream effects on many different insects and animals. For example, animals that burrow need soil that is loose enough for excavation, but strong enough to prevent a subsequent collapse. Arthropods, nematodes, mites, termites, and an array of desert insects rely upon the soil and its crust for food, shelter, and water, and many plants native to the Sonoran Desert rely upon burrowing animals and insects to create openings for seed growth and species abundance. For example, McAuliffe (1994) notes that creosote bushes benefit from growing on the mound systems created by kangaroo rats.

---

[1] For readers unfamiliar with the desert southwest, there are 17 different types of rattlesnakes that live in Arizona. The mountain preserves in/around Phoenix are loaded with Western Diamondbacks, Southwestern Speckled, Northern Blacktails, Tiger and Mojave rattlesnakes, and even Sonoran Sidewinders (also poisonous).

9.3 Urban/Wildland Interface

The Phoenix Mountains Preserves exist in a landscape dominated by humans. As a result, the preserves are a product of complex interaction between natural and human factors. For the purposes of this chapter, we focus on the Piestewa Peak preserve unit, located in North Central Phoenix (Fig. 9.2). Approximately 6000 people per week use the trails in/around the peak. There are several notable landmarks in Fig. 9.2 worth noting. First, Arizona State Route 51 (Piestewa Freeway) transects the preserve. Approximately 145,000 vehicles pass by the preserve, per day (AZDOT 2017). Second, there are three major access points to the preserve, including the Dreamy Draw Recreation Area, Piestewa Peak Park, and the 40th Street Trailhead—all of which are highlighted in Fig. 9.2. It is also important to note that there are other access points scattered throughout the area (not highlighted) that may consist of one or two parking places, with a short walk required to reach the trail system. Perhaps the most important feature for readers to notice from Fig. 9.2 is the residential and commercial development surrounding the preserve. It is completely enclosed by low- and medium-density residential developments and a handful of commercial corridors. This urban/wildland interface means that the

Fig. 9.2 Piestewa Peak preserve unit, Phoenix, Arizona

preserve maintains significant exposure to (and opportunities for) invasion by non-indigenous organisms, soil movement and compaction, air pollution, and again, massive levels of recreational use from people, pets, and pack animals.

Because the Piestewa Peak preserve area is so popular, there are significant access problems on weekends and holidays, where the formal access points for the preserve are overflowing with cars and hikers. During these times of elevated use, many potential park patrons are forced to find alternative points of access to the trail systems. As detailed previously, there are several smaller access points throughout the region, but those too become crowded during peak visitation times. Over time, this pushes potential visitors into scenarios where they are seeking alternative points of entry to the preserve. For example, consider Fig. 9.3, which highlights a

Fig. 9.3 Ad hoc access point, Piestewa Peak, Phoenix Mountains Preserve

residential cul-de-sac bordering the preserve area. Although portions of the preserve are fenced, there are many areas with fence failures (or no fence at all). The boulder pile/barrier illustrated in Fig. 9.3 is one way that park officials attempt to deal with lack of fencing and ad hoc access points created by visitors. This is one (of many) geographic points around the preserve where the urban/wildland interface occurs, directly.

The problem with these types of ad hoc access points include the many downstream effects on the preserve, itself. Efforts to connect to the formal trail system to these access locations generate a series of informal trails throughout the preserve (see Fig. 9.3). While some are relatively short, others are not, extending several hundred feet (or more) from the access points. Consider, for instance, Fig. 9.4a, which highlights the *formal* trail system for Piestewa. There are approximately 33.34 miles of mountain trails within the preserve, which includes portions of a large connecting trail (Charles M. Christiansen) to the North Mountain/Shaw Butte Preserve area, but does not include the multi-use trails (i.e., paved), for bicyclists and walkers.

Figure 9.4b provides a closer look at part of the trail system near the north edge of Piestewa. Of note are the many informal trails that exist in the landscape, most running in a north/south direction from the formal trail system. Although Fig. 9.4b is not pushing the limits of resolution for the composite LandSat 8 imagery, one can easily miss the nuances of the informal trail system with lower resolution satellite images. Recall that LandSat 8 provides a pixel resolution of 15 m for its panchromatic images. However, many of the informal trails are no more than 1 m wide at any given point. Figure 9.5 highlights one example of this. Specifically, Fig. 9.5a is the LandSat 8 image of a portion of our study area. The highlight box is overlaying an informal trail that is undetectable with the LandSat 8 imagery. Further, "zooming in" closer will not help in identifying the trail, it simply pixelates the area within the highlight box. However, Fig. 9.5b displays UAS-derived imagery within the highlight box. In this instance, one can easily detect the informal trail with the higher resolution imagery.

To reiterate, an outstanding challenge for environmental managers is the process of identifying, assessing, and interdicting the development of informal trails in sensitive landscapes. As detailed previously, the three most common approaches include *estimates*, *field inventories*, and/or *remote sensing*. When estimates and field inventories prove too time consuming, inaccurate and/or expensive, or when the spatial resolution of satellite images is too coarse for informal trail detection, unmanned aerial systems provide an alternative remote sensing platform for detecting informal trails. In the next section, we provide a more thorough case study for detecting informal trails in the Piestewa Peak preserve unit with UAS-derived imagery.

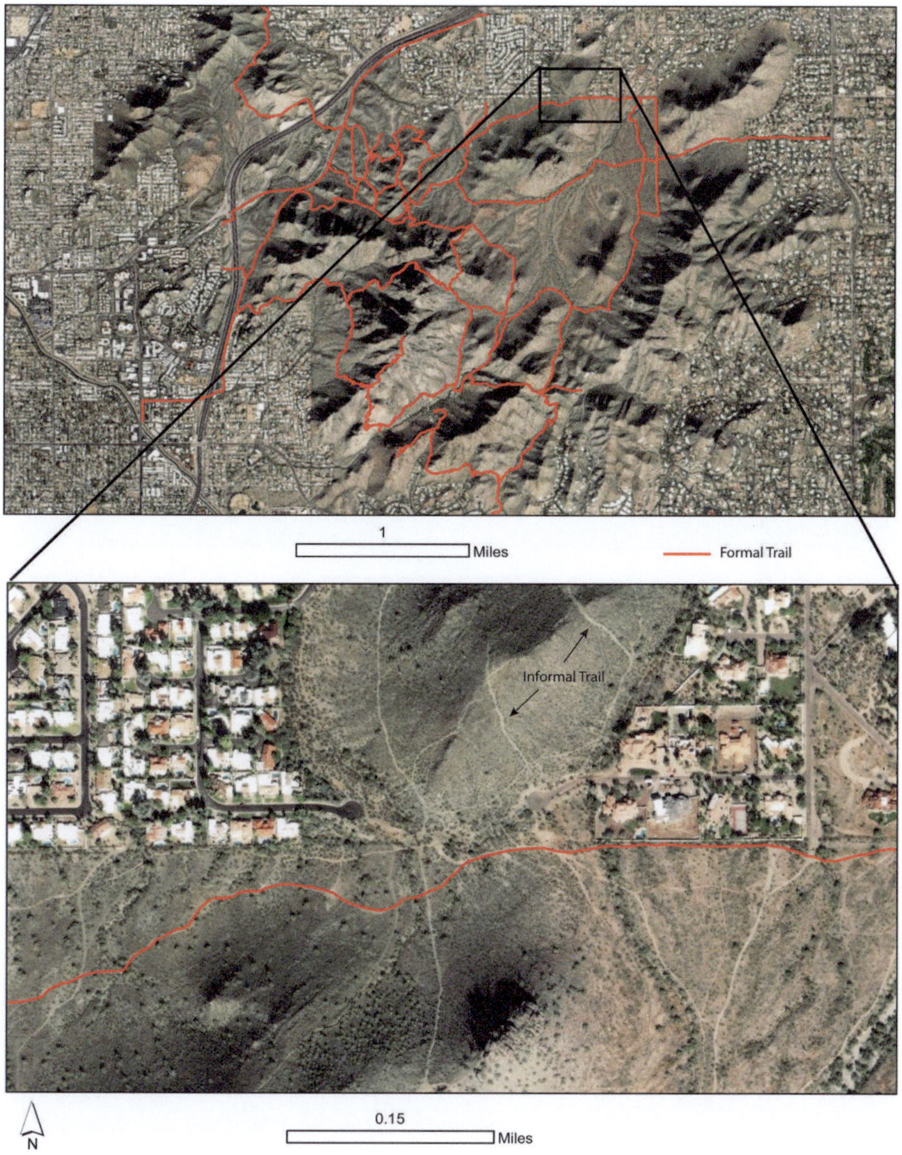

Fig. 9.4 Landsat8 imagery of Piestewa

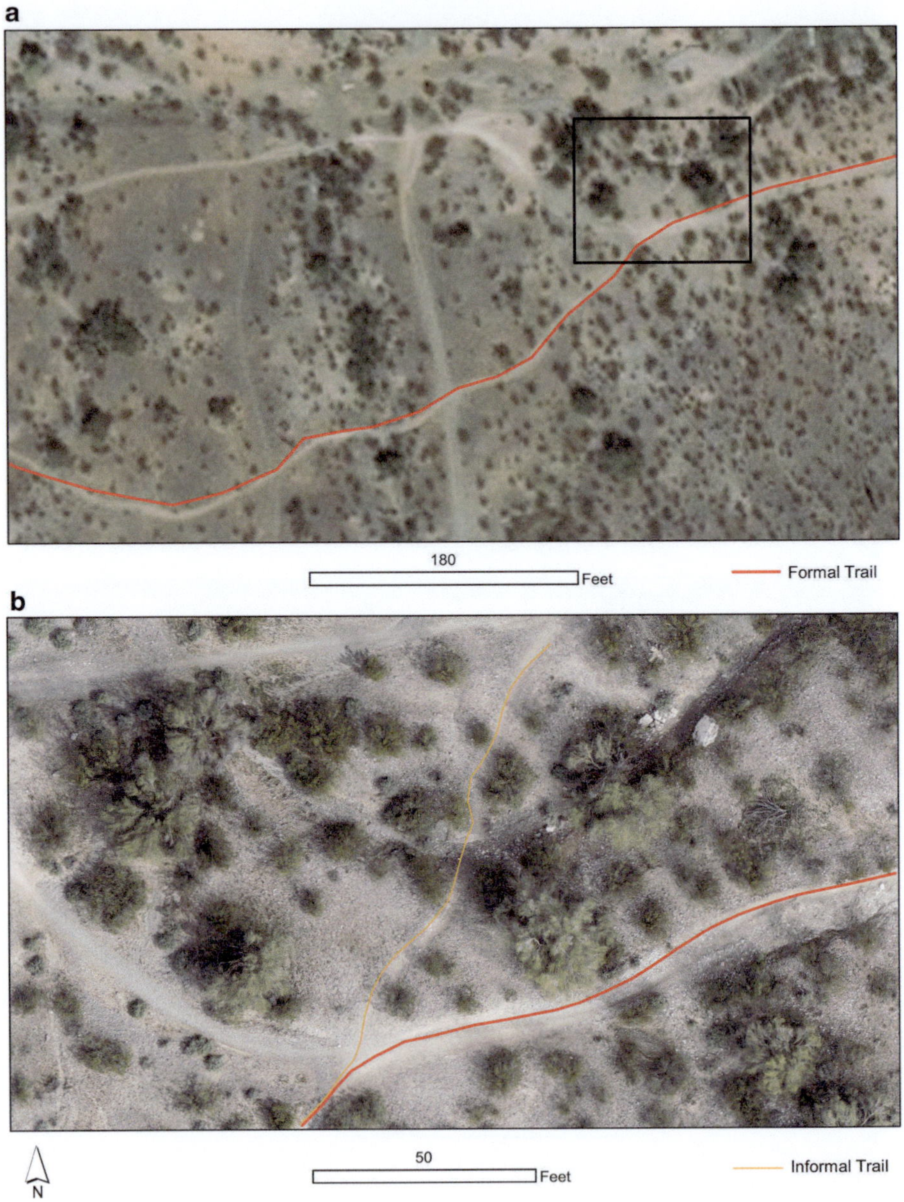

Fig. 9.5 (**a**) Landsat8 imagery, (**b**) sUAS imagery

9.4 Informal Trail Detection and Analysis with sUAS Imagery

9.4.1 Mission Details and Study Area

We flew a portion of the Piestewa unit of the Phoenix Mountains Preserves on March 28, 2019 around 10 am with the WingtraOne sUAS. As detailed throughout the book, the Wingtra is equipped with a Sony RX1RII 42-megapixel camera. This mission covered 84.18 ha (approximately 208 acres), with about 40 min of time airborne. We collected 832 geotagged images and achieved a ground sampling distance of 1.74 cm/pixel. We used Pix4D to process the imagery on a desktop computer with an Intel Core 7 processor running at 2.7 GHz with 64GB of RAM. Initial processing required 41 min and the orthomosaic required 3 h and 37 min of computing time. For readers keeping score, yes, a 3D point cloud was generated, which required almost 11 h of computational effort, yielding 264,524,589 points with an average density of 650 points per cubic meter.

We focus our analysis on a sub-area of the preserve, proximal to East Dunlap Avenue (Fig. 9.6). Consisting of approximately 116 acres, the analysis area is bordered by a low/medium density residential area to the north. More importantly, East Dunlap Ave is littered with a handful of ad hoc access points to the preserve (highlighted by blue squares in Fig. 9.6a). Again, these ad hoc entries are popular on weekends and holidays, when formal entries to the park are overflowing with vehicles and no local parking is available. Many neighborhood residents also use these access points, rather than driving to the formal entries, because it is the closest access to the preserve from their home. Unfortunately, it is these locations that contribute to the development and expansion of informal trails. Because the ad hoc access points are not connected to the formal trail system, visitors of the preserve create informal trails to either access established trails or connect with additional informal trails for enjoying the open space. One can clearly see these informal trails throughout Fig. 9.6. Of particular interest is Fig. 9.6b, which highlights one of the East Dunlap cul-de-sacs, its ad hoc access points, and the boulder barriers ringing the cul-de-sac. Again, while the boulders are effective in keeping motorized vehicles from entering the preserve, the interstitial spaces between boulders allow for the free flow of pedestrians, cyclists, and pack animals.

9.4.2 Software, Data Preparation, and Analytics

In addition to the image processing with Pix4D detailed above, we performed the remaining analysis in ArcGIS 10.7.1, a commercial geographic information system and FRAGSTATS 4.2, a spatial pattern analysis program for categorical data (ESRI 2019; McGarigal 2015).

Fig. 9.6 (**a**) Ad hoc access points in the Piestewa Preserve Unit, (**b**) Physical barriers and ad hoc access points

The first step in this analysis was to digitize both formal and informal trails in the study area. This includes the creation of polylines for each trail, as well as a series of polyline attribute categories for capturing the qualitative characteristics of the trails. Specifically, multiple width measurements were taken at equal intervals along each path. These measurements were averaged to obtain a representative path width. Based on these measurements, we created three categories of trails through the

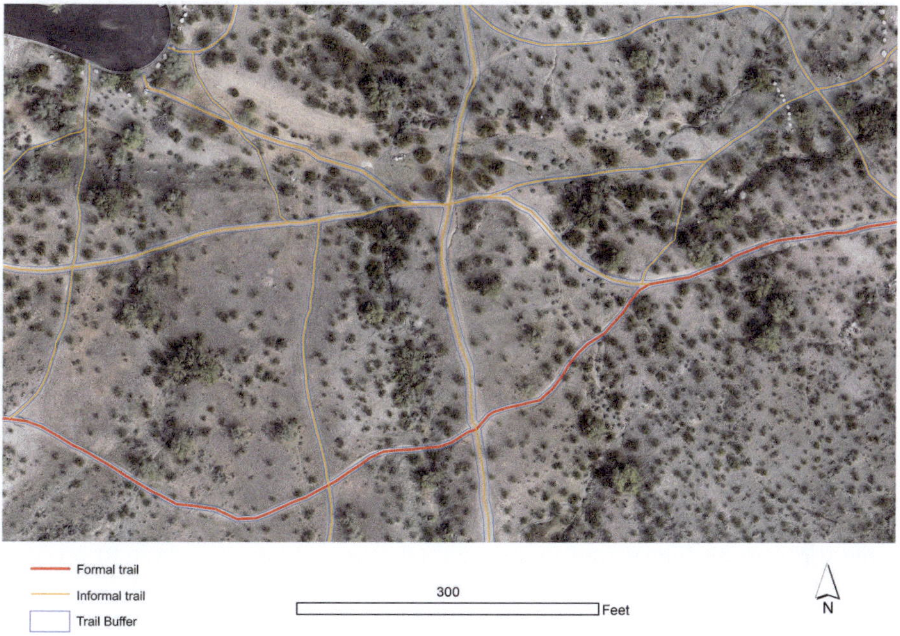

Fig. 9.7 Trail buffers

preserve: (1) *formal*, (2) *informal 1*, and (3) *informal 2*. As detailed previously, *formal* trails are maintained pathways through the preserve that are officially recognized by the city of Phoenix. For our purposes, *informal 1* trails represent larger, unofficial pathways created by hikers, bikers, and pack animal activity. In this instance, the trails averaged between 1.5 and 2.1 m across and displayed clear signs of heavy traffic. The second group of trails, *informal 2*, are single tracks through the desert landscape that displayed less foot and bike traffic and averaged between 0.3 and 0.8 m in width. Interestingly, only a single trail fell between these two groups, with an average width between 1.2 and 1.5 m. Ultimately, we classified this as an *informal 1*.[2]

Once each polyline was digitized and width measurements were attributed, we used a simple buffer operation in ArcGIS to generate a more realistic footprint of each trail in the preserve (Fig. 9.7). One key technical note regarding the generation of these buffers: if one selects the "full" option for the buffer in ArcGIS, it is critical that one also divides the width attribute by a value of 2. This is important because the "full" selection creates a buffer on either side of a polyline using the width

[2] It is important to note that there are likely many alternative strategies for capturing path widths. For example, if an extensive network of trails exists in a study area, one could pursue a random sample of trails within each category for generating both an average width and its standard deviation. In turn, one could use these measures to bind a distribution of random widths for each trail segment in the network.

attribute. Thus, if the database attribute is 2 m, one must divide this width by 2—which generates a 1 m buffer on each side of the polyline, yielding a 2 m wide trail.

In addition to the buffered trail polygons, we digitized polygon features for nearby residential areas to help delineate these areas from the preserve. For larger, more complex residential areas, one could also use a cadaster or parcel file for this purpose. Finally, both the trail and residential polygon features were assigned unique identifiers. Trail areas were assigned a value of 100 and residential areas a value of 200. Both were merged into a single shapefile, and the underlying values were used to create a raster coverage, with each cell inheriting a value of either 100 or 200. All remaining land in the study area was assigned a value of 300, corresponding to desert landcover. Finally, in addition to the full trail network, we created an additional landcover raster with *only* the formal trails to illustrate the effects of an informal trail network on landscape fragmentation.

For readers with limited training in ecological analysis, it is important to note that the resolution of landcover rasters can dramatically impact the computational effort required for yielding a viable analysis. For example, several of the more advanced FRAGSTATS tools evaluate the landscape *pixel by pixel*. This means that large (or even medium)-sized grids could take several hours or more to process. Recall that the sUAS-derived imagery for this analysis is 1.74 cm/pixel. To ease the computational burden without losing fidelity of analysis, we upscale the landcover raster to a spatial resolution of 1 ft/pixel. We then convert the grid to a Geotiff for analysis in FRAGSTATS.[3]

9.5 Results

One of the simplest metrics available for landscape analysis in the Piestewa preserve unit is a calculation of the geographic footprint of the formal and informal trail segments. Specifically, the "footprint" of the trail network is the total portion of the landscape that is covered by the trails. We calculate this by multiplying the length of the formal and informal trails by their width (Table 9.1). In this instance, the informal trail network is nearly eight times larger than the formal network.

Clearly, most of the trail network is not officially recognized by the preserve, and this reveals a significant disparity in trail structure (also use) in the preserve. However, the geographic footprint is only part of the story.

[3]FRAGSTATS can operate at different analysis levels and includes within patch metrics (Patch), between class metrics (Class), and landscape-level metrics (Landscape). Some of these metrics will only apply to certain analysis levels while others work across all levels. For a full list of FRAGSTAT metrics, we encourage readers to visit the FRAGSTATS website and to explore the accompanying overview report detailing the available methods (McGarigal and Marks 1995) and their broader context for ecological analysis (Cushman et al. 2008; Cushman et al. 2012). For the purposes of this analysis, we use both landscape- and class-level metrics.

Table 9.1 Descriptive statistics for the formal and informal trail network

Trail network	Total length (m)	Mean width (m)	Geographic footprint (m^2)
Formal	1051.11	1.37	1441.64
Formal + informal	7343.49	1.20	10,109.45

As previously detailed, disruptions to sensitive desert ecosystems can have long-lasting and far-reaching effects on the local flora and fauna, especially those sensitive to habitat fragmentation. Impacts vary for each species, but an ability to measure disruption to the landscape is a good first step in further quantifying how the impacts may manifest. The combination of sUAS imagery and FRAGSTATS makes possible a high-resolution analysis of disruptions.

9.5.1 Landscape-Level Results

One basic metric for quantifying fragmentation is the number of patches (NP) in the landscape. When using the formal trail system, a total of nine distinct patches are detected. When the informal trail system is included, 48 unique patches are detected (Table 9.2). This is a dramatic difference. However, it is important for readers to remember that there are no "real-world" scenarios associated with the formal trails only. The formal trails coexist with the informal trails in the preserve, no matter what. The use of metrics associated with formal trails, however, provide a good baseline for understanding what the preserve might be like if the informal trails did not exist or were interdicted and targeted for restoration efforts.

A second set of useful metrics focus on the edges themselves, including total length of edge (*TE*) and edge density (*ED*):

$$\mathrm{TE} = E \tag{9.1}$$

$$\mathrm{ED} = \frac{E}{A}\left(10{,}000\right) \tag{9.2}$$

where E is the total length (m) of edge in a landscape. If a landscape border is present, TE includes boundary segments representing the "true" edge only (i.e., abutting patches of different classes). If there is no landscape border, TE includes a user-specified proportion of the landscape boundary. A is the total landscape area in square meters. The interpretation of both measures is relatively simple. TE is the sum of the lengths (m) of all edge segments for a patch type. There is no upper limit on this value, but TE = 0 where there is no edge in the landscape. ED is a related measure that divides TE by the total landscape area (m^2), multiplied by 10,000 to convert to hectares.

We use FRAGSTATS to calculate both measures at the landscape-level. FRAGSTATS uses an exhaustive sampling technique via moving window with a search radius of 500 m. As detailed in Table 9.2, the total edge length and the edge

Table 9.2 Landscape-level metrics calculated with FRAGSTATS

Trail	NP	TE	ED	GYRATE_ AM	PROX_ AM_L	CONNECT	SPLIT
Formal	9	12,896	25.1783	794.1907	9170.7876	46.1538	1.6591
Formal + informal	48	61,784	120.6383	321.3538	8012.015	32.4487	9.2259

density are higher for the informal trail network, except for the residential landcover (RES) class which will not vary because the edge length does not change between the two landscapes. Regardless, the informal trail network creates high levels of edge density throughout the landscape and suggests that fragmentation may be a problem. In fact, one may use the area-weighted mean radius of gyration (GYRATE_AM) to determine how far an unsegmented habitat patch extends across a landscape. As noted by McGarigal (2015), when an organism is given a random starting point, the radius of gyration measures the average distance an organism can move within a patch before encountering a boundary. Not surprisingly, the informal trail system with Piestewa reduces the radius of gyration significantly. Similarly, the results of the FRAGSTATS proximity metric (PROX_AM_L), which measures the average distance required to reach another habitat patch, suggests that there is very little difference in the landscape when comparing the formal and informal trail systems.

Lastly, we use FRAGSTATS to calculate the *connectance index* (CONNECT) and the *split index* (SPLIT), both of which measure connectivity. Specifically, when the landscape is a single habitat or when the patches do not connect, CONNECT = 0. When every patch in the landscape connects (or patches of similar types when analyzing at the class-level), CONNECT = 100. Not surprisingly, Table 9.2 suggests more patch connectivity for the preserve if the formal network existed alone. The formal network has about 46% connectivity. However, with the emergence of the informal trail system, the connectivity drops to 32.44%. Conversely, SPLIT increases as the landscape becomes subdivided. When SPLIT = 1, the landscape is a single habitat patch. For the Piestewa preserve unit and its formal trail system, the SPLIT value is 1.67, suggesting little habitat subdivision. However, when the informal trails are included, the SPLIT score increases to 9.66, indicating a substantial increase in subdivision.

9.5.2 *Class-Level Results*

The class-level metrics for Piestewa echo those calculated for the landscape-level. In short, the informal trail system drastically inflates the total number of edges and edge density for the desert habitat. All the values for the radius of gyration, proximity, connectance, and split metrics (Table 9.3) suggest that informal trails fragment the preserve.

Table 9.3 Class-level (habitat) metrics calculated for both the formal and informal trail networks

Trail	CLASS	NP	TE	ED	GYRATE_AM	PROX_AM_C	CLUMPY	CONNECT	SPLIT
Formal	RES	5	4504	8.79	235.88	4.12	0.99	40.00	308.83
Formal	DES	3	12,896	25.18	845.34	10,074.0	0.97	66.67	1.67
Formal	TRL	1	8392	16.38	807.01	0.00	0.75	0.00	26458.6
Informal	RES	4	4504	8.67	236.18	9.38	0.99	33.33	309.69
Informal	DES	41	61,724	120.52	302.10	9153.76	0.89	32.20	9.66
Informal	TRL	3	57,404	112.09	899.91	0.13	0.72	100.00	603.74

A final measure, the *clumpiness index* (CLUMPY), is a class-level only standardized metric with values that vary between −1 and 1. CLUMPY measures the overall level of aggregation between patches of a specified habitat class. Values approaching 1 suggest that patches of a habitat class are aggregated when compared to a spatially random distribution. Values in/around 0 suggest that patches are randomly distributed across the landscape. Values approaching −1 suggest greater dispersion than expected versus a spatial random distribution. Once again, when focusing on the desert habitat class as an example, the inclusion of formal trails results in a CLUMPY value of 0.97, but when one includes informal trails too, CLUMPY decreases to 0.89. This suggests that the informal trail system disaggregates the desert habitat.

9.6 Discussion

There are several results of the analysis worth further discussion. First, the vast majority of the trails detected by the sUAS in our study area are informal. This is not because Piestewa is lacking formal trails or access points to them. Rather, it suggests that visitors are using large portions of the preserve more informally, adding mechanical disturbances to an already sensitive ecosystem. One of the key factors fueling this use is the lack of meaningful signage within the park. Figure 9.8 highlights the most frequent sign encountered in the preserve area which alerts visitors that no motorized vehicles are allowed within the park. While this is an important notification, there is no guidance on these signs regarding the responsible use of trails within the preserve. Thus, while all of the formal access points display signage encouraging visitors to tread lightly on the Sonoran landscape by not "trailblazing," it is unlikely that most locals (or even tourists) consult this information prior to a visit.[4]

A second point worth discussion is the sheer number of ad hoc access points in/around residential areas that border the preserve. In this case, the cul-de-sacs for both sides of East Dunlap Avenue offer informal access points to the preserve. There are seven ad hoc access points within 0.53 miles of each other in this residential area. As detailed previously, these types of access points represent a subset of the primary gateways to the informal trails scattered throughout the preserve. Additional efforts to build barriers (e.g., fencing/boulders) in/around the area may significantly reduce the use of several informal trails established over the years, facilitating landscape recovery. At the very least, additional fences and rock barriers can help reroute preserve visitors to a single informal trail, rather than visitors scattering to multiple informal trails for accessing the preserve's open space. The boulder barriers detected in the sUAS imagery are effective for keeping vehicles out of the

[4] This information is also available on several websites, including http://www.phoenixmountains.org

Fig. 9.8 Phoenix Mountains Preserve signage

preserve, but cyclists and pedestrians can easily travel around/over the boulders for accessing informal trails. As a result, fencing may be the best option for re-routing hikers and bikers, but this can be expensive.

Third, the geographic footprint of informal trails versus formal trails is revealing. Recall from the results that the formal trail system within the study area consists of 1051.11 m, but the informal trails span 7343.49 m of the study area. The city is certainly aware of some of these trails, which are labeled as "undesignated" in the official preserve map (Fig. 9.9); however, most of the informal trails do not appear on the official record. More importantly, emerging trails are difficult to detect using satellite imagery. Recall the resolution challenges associated with the LandSat 8 imagery (Fig. 9.5). With a resolution of 15 m, identifying new informal trails is nearly impossible. Again, this is why the sUAS imagery provides such a powerful tool for environmental managers. With a resolution of approximately 1.74 cm/pixel, the sUAS-derived images provide an extraordinary window into the Piestewa preserve unit and its informal trail system (even when upscaled to 1 m). Again, these trails represent the most significant disruptive force to the desert preserve,

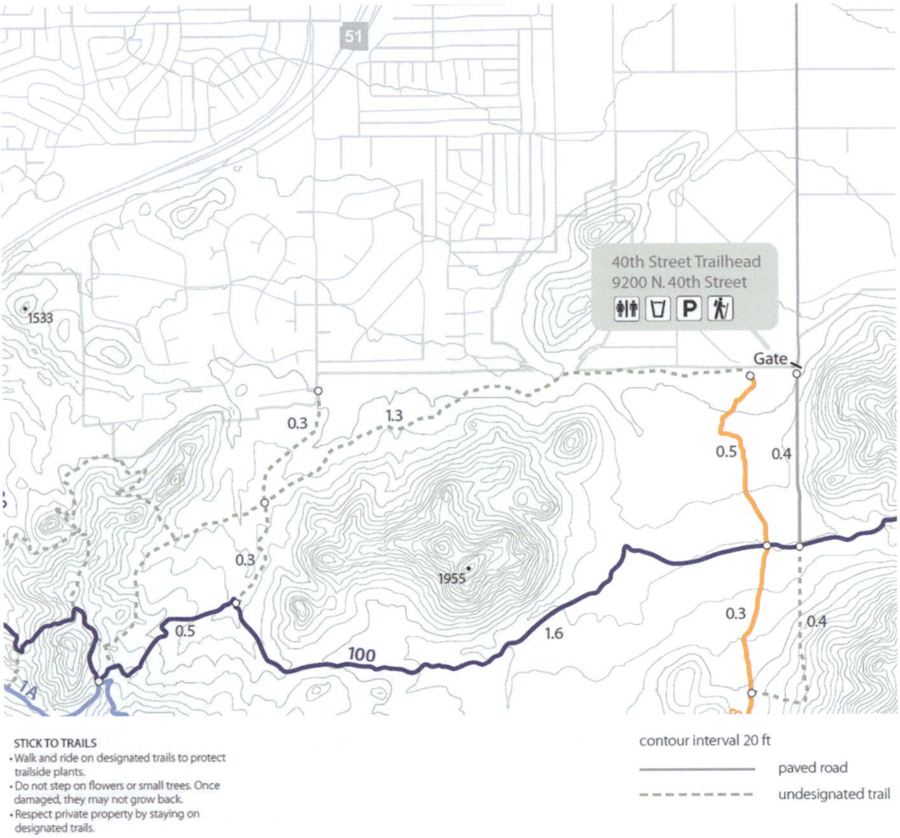

Fig. 9.9 A portion of the Phoenix Mountains Preserve trail map, 2019. (Source: City of Phoenix (https://tinyurl.com/y3lyvspo))

fragmenting the landscape, contributing to the mechanical disruption of sensitive soils and providing visitors access to the most pristine sections of the preserve—all largely independent of the formal (i.e., managed) trails. In addition to limiting ad hoc access points, it is possible to develop and implement some habitat restoration strategies throughout the preserve. Lovich and Bainbridge (1999) review several restoration efforts in desert regions throughout the United States, primarily focusing on revegetation. The results are not particularly promising. Plant growth and establishment in extreme desert environments are slow, and efforts to revegetate disturbed areas frequently suffer from subsequent mechanical disturbances such as grazing (e.g., livestock and/or wildland fauna, including rabbits) and severe meteorological events (e.g., monsoon rains). However, the most successful efforts included the use of native flora, protection from wildland fauna, and periodic irrigation. Given the geographic expanse of Piestewa, this may not be realistic, but efforts to interdict the use of informal paths with boulders or some other type of obstruction may help reduce the wear-and-tear caused by hikers, bikers, and pack animals.

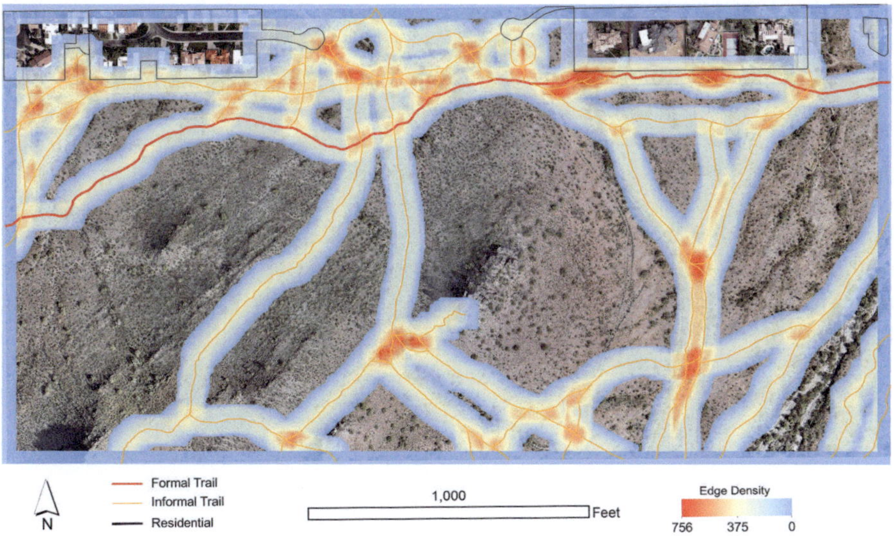

Fig. 9.10 Informal trail edge density

Finally, the landscape fragmentation within the Piestewa preserve unit is severe. The combination of informal trail length, width, and density helps to fragment the landscape in profound ways. For perspective, Fig. 9.10 displays the edge density of the study area and reinforces just how badly the informal trails carve up the desert landscape. Tables 9.2 and 9.3 statistically support this conclusion. Perhaps most instructive are the results for the area-weighted mean radius of gyration measure, which suggests that the informal trail system reduces the average distance an organism can move within a Piestewa desert patch before encountering a boundary from 845 to 302 m.

9.7 Conclusion

The emergence of informal trails in parks, wilderness areas, and protected natural areas continues to be a significant problem. The ecological impacts can be stark, ranging from mechanical disturbances that impact sensitive vegetation and the introduction of invasive species to soil compaction and its corresponding reduction in water holding capacity and changes to precipitation run-off patterns. Although stewards of these sensitive areas can use a variety of methods for detecting and cataloging the informal trails, including *estimates*, *field inventories*, and *remote sensing*, there are limitations to these approaches. While estimate-based methods are viable for identifying the presence of trails, very little information is generated regarding trail characteristics (e.g., length, width, density). Comprehensive field inventories provide significantly more information, but are time consuming and expensive and can be dangerous. Finally, remote sensing methods can provide a more panoptic view of informal trails in management area, but satellite imagery can suffer from poor resolution, especially when the informal trails are small in footprint (e.g., width) or recently emergent—like many of those in the Piestewa preserve unit. This chapter highlights how remote sensing via small unmanned aerial systems provides an excellent foundation from which to build a better snapshot of informal trail networks, their geographic footprint, and their potential ecological impacts. It is also important to note that although this chapter was limited to a single snapshot in time, it is possible to collect a time-series database of the study area by simply flying additional drone missions. With these data, one could easily track changes to the trail system over time, identifying changes to the landscape with minimal effort.

References

Arizona Department of Transportation [AZDOT]. (2017). *AADT reports: Traffic counts*. Retrieved August 17, 2019, from https://tinyurl.com/yxby4p6m

Barros, A., & Pickering, C. M. (2017). How networks of informal trails cause landscape level damage to vegetation. *Environmental Management, 60*(1), 57–68.

Boorman, L. A., & Fuller, R. M. (1977). Studies on the impact of paths on the dune vegetation at Winterton, Norfolk, England. *Biological Conservation, 12*(3), 203–216.

Cushman, S. A., McGarigal, K., & Neel, M. C. (2008). Parsimony in landscape metrics: Strength, universality, and consistency. *Ecological Indicators, 8*(5), 691–703.

Cushman, S. A., Shirk, A., & Landguth, E. L. (2012). Separating the effects of habitat area, fragmentation and matrix resistance on genetic differentiation in complex landscapes. *Landscape ecology, 27*(3), 369–380.

Dale, M. (2018). Is south mountain park the country's largest? *KJZZ*. Retrieved August 12, 2019, from https://tinyurl.com/y55g7fsp

ESRI. (2019). *ArcGIS desktop: Release 10.7.1*. Redlands, CA: Environmental Systems Research Institute.

Forman, R. T. T. (1995). *Land mosaics*. New York: Cambridge University Press.

Forman, R. T., & Alexander, L. E. (1998). Roads and their major ecological effects. *Annual Review of Ecology and Systematics, 29*(1), 207–231.

Gallen, T. (2019). Phoenix leads US in population growth, new census data shows. *Phoenix Business Journal*. Retrieved August 19, 2019, from https://tinyurl.com/y5f3eb2j

Hammitt, W. E., & Cole, D. H. (1998). *Wildland Recreation: Ecology and Management*, 2nd ed. John Wiley &Sons, New York.

Hammitt, W. E., Cole, D. N., & Monz, C. A. (2015). *Wildland recreation: Ecology and management*. Hoboken: Wiley.

Hawes, A. (2005). Surviving the sonoran. *Zoogoer*. Retrieved August 25, 2019, from https://tinyurl.com/yxrnccmp

Jägerbrand, A. K., & Alatalo, J. M. (2015). Effects of human trampling on abundance and diversity of vascular plants, bryophytes and lichens in alpine heath vegetation, Northern Sweden. *Springerplus, 4*(1), 95.

Kiefer, M. (1998). Deconstructing the Phoenix Mountain Preserve. *Phoenix New Times*. Retrieved August 16, 2019, from https://tinyurl.com/y4syza4o

Korpilo, S., Virtanen, T., & Lehvävirta, S. (2017). Smartphone GPS tracking—Inexpensive and efficient data collection on recreational movement. *Landscape and Urban Planning, 157*, 608–617.

Kutiel, P. (1999). Tendencies in the development of tracks in open areas. *Environmental Management, 23*(3), 401–408.

Landinfo. (2018). *Prices*. Retrieved August 14, 2019, from http://www.landinfo.com/prices.htm

Leung, Y. F., Newburger, T., Jones, M., Kuhn, B., & Woiderski, B. (2011). Developing a monitoring protocol for visitor-created informal trails in Yosemite National Park, USA. *Environmental Management, 47*(1), 93–106.

Lovich, J. E., & Bainbridge, D. (1999). Anthropogenic degradation of the southern California desert ecosystem and prospects for natural recovery and restoration. *Environmental Management, 24*(3), 309–326.

Marion, J. L., Leung, Y. F., & Nepal, S. K. (2006). Monitoring trail conditions: New methodological considerations. In *The George Wright forum* (Vol. 23, pp. 36–49). George Wright Society.

Martin, R. H., Butler, D. R., & Klier, J. (2018). The influence of tire size on bicycle impacts to soil and vegetation. *Journal of Outdoor Recreation and Tourism, 24*, 52–58.

McAuliffe, J. R. (1994). Landscape evolution, soil formation, and ecological patterns and processes in Sonoran Desert bajadas. *Ecological Monographs, 64*(2), 111–148.

McGarigal, K. (2015). *FRAGSTATS*. URL: https://www.umass.edu/landeco/research/fragstats/fragstats.html

McGarigal, K., & Marks, B. J. (1995). *FRAGSTATS: spatial pattern analysis program for quantifying landscape structure*. Gen. Tech. Rep. PNW-GTR-351. Portland, OR: US Department of Agriculture, Forest Service, Pacific Northwest Research Station. 122 p, 351.

McGarigal, K., & Romme, W. H. (2012). Modeling historical range of variability at a range of scales: An example application. In J. Wiens, G. Hayward, H. Safford, & C. Giffen (Eds.), *Historical environmental variation in conservation and natural resource management* (pp. 128–146). Hoboken: Wiley.

McGivney, A. (2019). Yanked from the ground': Cactus theft is ravaging the American desert. *The Guardian*. Retrieved August 12, 2019,from https://tinyurl.com/y2hk6r4l

Odegard, K. (2014). Forest crime: Thieves steal plans, moss, fungi, damage environment. *Albany Democrat-Herald*. Retrieved August 15, 2019, from https://tinyurl.com/y6ep5oye

Pescott, O. L., & Stewart, G. B. (2014). Assessing the impact of human trampling on vegetation: A systematic review and meta-analysis of experimental evidence. *PeerJ, 2*, e360.

Pickering, C. M., & Growcock, A. J. (2009). Impacts of experimental trampling on tall alpine herbfields and subalpine grasslands in the Australian Alps. *Journal of Environmental Management, 91*(2), 532–540.

Pickering, C. M., & Norman, P. (2017). Comparing impacts between formal and informal recreational trails. *Journal of Environmental Management, 193*, 270–279.

Rothwell, S. (2014). Background of acquisition of land for the Phoenix Mountain Preserves and how Phoenix Mountain Preservation Council (PMPC) was able to influence the land acquisition program. *Phoenix Mountains Preservation Council*. Retrieved August 27, 2019, from https://tinyurl.com/yxamlj7a

Wahl, J. (2016). Man rescued after hiking North Mountain with beer, no water. *12 News*. Retrieved August 12, 2019, from https://tinyurl.com/y696u9ys

Walden-Schreiner, C., & Leung, Y. F. (2013). Spatially characterizing visitor use and its association with informal trails in Yosemite Valley meadows. *Environmental Management, 52*(1), 163–178.

Westcott, F., & Andrew, M. E. (2015). Spatial and environmental patterns of off-road vehicle recreation in a semi-arid woodland. *Applied Geography, 62*, 97–106.

Willard, B. E., Cooper, D. J., & Forbes, B. C. (2007). Natural regeneration of alpine tundra vegetation after human trampling: A 42-year data set from Rocky Mountain National Park, Colorado, USA. *Arctic, Antarctic, and Alpine Research, 39*(1), 177–183.

Wimpey, J., & Marion, J. L. (2011). A spatial exploration of informal trail networks within Great Falls Park, VA. *Journal of Environmental Management, 92*(3), 1012–1022a.

Chapter 10
Drone Futures

Abstract The future is exceedingly difficult to predict, especially when predictions focus upon newly emerging, disruptive technologies such as unmanned aerial systems. Nevertheless, the purpose of this chapter is to provide a prospective look at the future of unmanned aerial systems for urban spatial analysis. This includes challenges associated with the rapid advancements in technological platforms and applications. We also briefly address the potential challenges associated with the social impacts of drone technology for spatial analysis and the need for coherent policies to manage sUAS activity in urban environments.

Keywords Sputnik · Luna · Confirmation bias · Control autonomy · Miniaturization · Low power · Accelerators · Hardware · Computation · Static · Dynamic · Yoda

Prediction is very difficult, especially if it's about the future.
—Nils Boer

10.1 Introduction

Nils Boer is correct. Prediction is difficult. We know this because the past is littered with thousands of erroneous predictions about the future. Here are a few classically bad examples from people who are a lot smarter than the authors of this book:

> *Fooling around with alternating current is just a waste of time. Nobody will use it, ever.*
> Thomas Edison (1889)

> *There is not the slightest indication that nuclear energy will ever be obtainable. It would mean that the atom would have to be shattered at will.*
> Albert Einstein (1932)

> *We will never make a 32-bit operating system.*
> Bill Gates (1989)

T. H. Grubesic, J. R. Nelson, *UAVs and Urban Spatial Analysis*,
https://doi.org/10.1007/978-3-030-35865-5_10

There is practically no chance communications space satellites will be used to provide better telephone, telegraph, television, or radio service inside the United States.
T.A.M. Craven (1961)

The quote attributed to Craven, the former commissioner of the Federal Communications Commission, is an interesting one for many reasons. To put the quote into historical context, it is important to remember that human efforts to reach space were just beginning in the late 1950s and early 1960s. For example, the Sputnik 1 satellite from the USSR achieved a low earth orbit in October 1957. The US response to the success of Sputnik 1 was the Vanguard 1, which suffered a launchpad explosion during the same year. Although the US managed to get Explorer 1 and SCORE (a small communications satellite) into earth's orbit in 1958, it was the USSR launch of the Luna 1 and Luna 2, each of which contained basic radio equipment such as a tracking transmitter and telemetry system, as well as a series of sensors that included a magnetometer, Geiger counter, scintillation counter, and a micrometeorite detector (NASA 2019) that captured the world's imagination. The mission goal for Luna 1 was to collect measurements on magnetic fields, cosmic rays, and meteoric particles before it crashed into the moon. However, because of programming error, a rocket burn phase misfired and Luna 1 missed the moon completely. In September of the same year, the USSR launched Luna 2, which collected important measurements and crashed into the moon, as planned.

Although there were some additional developments associated with the space race in 1960 and 1961, including Yuri Gagarin's 108 min flight on Vostok 1 (completing a single orbit of the earth), Craven's skepticism was not misplaced. Not only was the successful launch of a satellite extremely difficult, the satellites themselves were fairly rudimentary, completely incapable of handling large-scale broadcast communications. However, in July 1962, NASA launched Telstar 1 into a low-earth orbit, where it beamed the first television and telephone calls. Then, in 1965, Intelsat I (aka, Early Bird) was the first commercial communication satellite to be placed in geosynchronous orbit.

How could Craven miss the mark so badly? Literally, 4 years after he made his prediction, people are watching television via satellite. Cognitive biases likely played a role in Craven's poor prediction, as well as those from Gates, Einstein, and Edison. Cognitive biases take many forms, including confirmation bias,[1] anchoring bias,[2] and the misinformation effect.[3] There are others. For a gentle overview, see Cherry (2019) and Morgan (2019) and for a deeper dive on additional biases, see Fischhoff (1975) on hindsight bias and Weinstein (1982, 1989) on optimism bias.

There is no doubt that the authors of this book suffer from every cognitive bias known to man but that will not keep us providing a prospective look at the future of

[1]Confirmation bias allows people to favor information that conforms to their existing beliefs, discounting evidence that does not conform (Nickerson 1998).

[2]Anchoring bias allows people to rely too heavily on the first pieces of information they learn (Furnham and Boo 2011).

[3]The misinformation effect allows post-event information to interfere with a person's memory of the original event (Ayers and Reder 1998).

unmanned aerial systems for urban spatial analysis. We draw upon the work of Floreano and Wood (2015), who made some very good predictions about drones in 2015, many of which remain relevant today. Specifically, we address several emerging challenges associated with the rapid advancements in UAS platforms and their applications. We will also briefly address the potential challenges associated with the social impacts of drone technology for spatial analysis and the need for coherent policies to manage UAS activity in urban environments.

10.2 The Future of sUAS Technology

There are several relatively easy predictions to make concerning the future of sUAS technology—not because they are completely obvious, rather because they are persistent, long-term challenges that need to be addressed before sUAS platforms transform. Specifically, three trends are worth watching: (1) *control autonomy*, (2) *miniaturization*, and (3) *low-power hardware accelerators*.

10.2.1 Control Autonomy

As detailed by Floreano and Wood (2015) and the International Organization for Standardization (ISO 2012), robot autonomy refers to the ability of a device and/or platform (e.g., UAS) to perform intended tasks based on its current state (and sensing) without human intervention. Consider the following example. If one is flying an sUAS at long distances and battery power is beginning to reach levels where the sUAS cannot fully complete its mission and return home safely, can the sUAS decide to temporarily abort its mission to obtain a fresh charge at a remote station and then resume its mission when battery levels meet an acceptable threshold? If time is of the essence, a full charge many not be necessary. Instead, the sUAS could accept a charge just large enough to finish the mission and return to the home waypoint for a safe landing. This may seem like a relatively simple task, but it requires copious amounts of information and computational processing. This includes data concerning the energy requirements for completing sensing operations, calculating distances to a remote charging station, flight distances required to complete the sensing mission and travel home, ambient environmental conditions (e.g., wind) that may impact flight longevity and many related tasks.

The ability of an sUAS to make these types of decisions rests on three levels of increasing control autonomy (Floreano and Wood 2015): (a) *sensory-motor autonomy*, (b) *reactive autonomy*, and (c) *cognitive autonomy*. Sensory-motor autonomy includes the ability for an sUAS to translate high-level human commands to perform tasks (this is typically accomplished via the flight planning software). Tasks may include the sUAS reaching and operating at a particular altitude, using coordinates from global positioning systems to perform the mapping/sensing of a target area, and making sure the sUAS can accommodate and use a large range of platform-dependent control signals (e.g., maintaining speed, appropriate pitch, approach angles) during a mission. Sensory-motor autonomy already exists in most mid-grade (or better) sUAS. Reactive autonomy (which requires sensory-motor autonomy) includes sUAS abilities to maintain its current position or trajectory when external perturbations are present. The perturbations might include wind, rain,[4] radio interference, electro-mechanical failure, stationary obstacles (e.g., building or wires), and moving obstacles (e.g., drones, birds), along with an ability to takeoff, land and/or coordinate with other moving objects (e.g., drone swarm). Many of the more advanced sUAS platforms, including the Sensefly Albris (2019), have advanced situational awareness sensors on board to facilitate reactive autonomy. Finally, cognitive autonomy (which requires reactive autonomy) allows the sUAS to perform simultaneous localization and mapping (SLAM),[5] resolve conflicting information about mission goals and conditions, recognize and avoid sensitive objects (e.g., people), and an ability to learn. At the time this chapter was written (September 2019), there are no commercially available sUAS systems with these capabilities. In part, this is because the amount of hardware required to perform such tasks quickly overwhelm smaller platforms with multiple digital cameras (increasing weight) and computationally expensive algorithms (which also require more onboard computational units, adding weight).

Control autonomy is critically important for urban areas. For example, Mohamed et al. (2018) detail many of the potential applications of UAVs in smart cities, from traffic monitoring and emergency health services, to security, environmental monitoring, tourism support, commerce, and communication. The lack of control autonomy for sUAS in just one of these sectors would be disastrous. However, imagine the needs for control autonomy when all these sectors are using sUAS simultaneously in urban skies. In short, movement toward cognitive autonomy for sUAS is inevitable if urban areas (and their associated public and private entities) plan on integrating drone technology, en masse, across multiple sectors.

10.2.2 Miniaturization and Low-Power Hardware Accelerators

Miniaturization refers to the process of making elements of a device smaller, reducing its weight and/or volume. Examples of miniaturization abound and some of them are really amusing. Readers may recall the movie *Wall Street*, where the antagonist Gordon Gecko (played by Michael Douglas) is walking on the beach making a phone call on a Motorola DynaTAC 8000×. Weighing 1.75 lbs (790 g), the phone

[4] As detailed previously, flying in the rain is a terrible idea.

[5] Recall that simultaneous localization and mapping (SLAM) refers to an "ability for a mobile robot (or UAS) be placed at an unknown location in an unknown environment and for the robot to incrementally build a consistent map of the environment while simultaneously determining its location within this map" (Durrant-Whyte and Bailey 2006, p. 99).

offered 30 min of talk time and required approximately 10 h to achieve a full battery charge (Ha 2010). Today, the Zanco Tiny T1 is available for users who are really concerned about mobile phone size. Not only does it make phone calls, it provides users with the ability to send text messages too. At a weight of 13 g, it is much smaller than the DynaTAC 8000× and way more useful (Saha 2017).

One can find examples of miniaturization in many places, from cell phones to portable power adapters, medical devices, and semiconductors (Peercy 2000). Miniaturization is also coming to unmanned aerial systems, but the physics are somewhat limiting. Floreano and Wood (2015) note that because flight is energetically expensive, reductions in sUAS size can be difficult to achieve. Specifically, as electromagnetic motors become smaller and less dense, energy transmissions are less efficient, and there are greater viscous losses because of reduced Reynolds numbers (Floreano and Wood 2015). Basically, like water, air also has viscosity. Thus, when flying an sUAS, it needs to overcome both density (inertia) and viscosity (stickiness) because the air (like water) is reluctant to move aside for the sUAS. These basic challenges for actuation, power, scaling, and ultimately flight, when mixed with the need for onboard sensors and computation, mean that the sUAS miniaturization process requires careful system-level design choices that balance flight needs with sensing needs. Also, as mentioned in the previous section, for an sUAS to achieve cognitive autonomy, the required number of sensors and onboard computational capabilities must increase. This adds weight, further taxing the smaller, less efficient electromagnetic motors for sUAS platforms.

One way to reduce the need for general computation onboard sUAS is the use of low-power hardware accelerators. For readers unfamiliar with this concept, the idea is a simple one. Specialized computing hardware, structured to perform a very specific task or set of tasks, can do this more efficiently than software running on a general purpose central processing unit (CPU) (Intel 2008). For example, desktop and laptop computers use general purpose CPUs to perform thousands of tasks, from word processing to mapping. However, any data transformation, calculation, or computing routine that can be accomplished with software and a CPU can also be accomplished on specialized hardware. For example, desktop computers often use (hardware accelerated) graphics cards to ensure that graphic-intensive applications (e.g., video games) are rendered quickly, with quality. This is not because the CPU is incapable of handling the task(s) of rendering graphics, but offloading these tasks to a specialized hardware element makes everything run more smoothly, decreasing the overall computational load to the CPU.

This is a critically important concept when considering efforts to miniaturize sUAS. For example, rather than including a general purpose CPU and a host of sensors controlled by software through the CPU, one could develop specialized hardware for performing important sensing tasks. This would likely include dedicated hardware for operating a multi-layer convolutional neural network for obstacle recognition (Ovtcharov et al. 2015), hardware that manages energy consumption, specialized hardware for thermal sensing tasks, and hardware to handle many other operations required for achieving control autonomy.

One major challenge associated with adding all of this specialized hardware to an sUAS is power consumption. Micro UAS (μUAS) will have small batteries and small motors and will require equally small onboard hardware and sensors. Most important, however, will be the management of power consumption. Take, for example, a desktop computer running an 8 GB AMD Radeon RX 590 graphics card. The graphics card is rated at 225 W of power, but this does not mean it is always using that many watts. Most computers include an array of power conservation tools to minimize consumption when idle. However, if one is playing Overwatch or Rainbow Six Siege, the power consumption of the graphics card will move toward the 225 W.

Things are no different for a μUAS. When deployed and collecting data in the field, all of the hardware will require power from the onboard supply. Thus, the development of low-power hardware accelerators is an absolutely critical step in the evolution of μUAS, as is the development of small, long-lasting rechargeable batteries (Armand and Tarascon 2008). Power conservation and management of hardware typically concerns both *static* and *dynamic* power. Static power is basically leakage current—the power consumed through a device when transistors are static. As detailed by Intel (2008), this is relatively constant for a given device. However, smaller devices typically consume less static power. Dynamic power is the energy consumed when current flows through transistors and they are actively switching states. Low-power hardware accelerators need to focus on reducing dynamic power use for the onboard sensors and their embedded computational units.

Research and development on low-power hardware accelerators has important implications for a number of other UAV capabilities. For example, the possibility of creating solar-powered UAVs becomes much more feasible when onboard systems reduce power consumption. At the moment, there are only a handful of sUAV utilizing solar power for flight. One example is the Puma AE UAV (https://tinyurl.com/y2gkt8wq) which flew more than 9 h using the energy generated by solar panels affixed to the top of the craft.[6] However, if solar-powered UAVs are going to do anything more than just fly (i.e., operation of active sensors), they will need to incorporate some form of low-power hardware acceleration mentioned above.

10.3 μUAS for Urban Spatial Analysis

Now that the basic physical limitations of sUAS and μUAS are established, at least for the moment, let us move a few years into the future and assume that these limitations no longer exist. Micro unmanned aerial systems are operating with full control autonomy, and low-power hardware acceleration allows for a suite of onboard sensing equipment from RGB cameras to multispectral and thermal sensors. Again, it is

[6] https://www.avinc.com/resources/press-releases/view/aerovironment-solar-powered-puma-ae-small-unmanned-aircraft-achieves-contin

Table 10.1 Internet data streams

Domain	Companies/platforms
Social/entertainment	Snapchat, Tinder, Life360, Periscope
IoT and smart home	Amazon Echo, Nest, LIFX
Connected vehicles	Uber, Lyft, Delphi, Easy Taxi, Gett
Financial and enterprise	Ameritrade, Seeking Alpha, SAP
Online collaboration	Slack, HipChat, Hive
Advertising/audience participation	Coca Cola, McDonalds, Yahoo

worth referencing a movie here. Readers might recall the μUAS named Redwing,[7] deployed by the character Falcon, in the Marvel movie, *Captain America: Civil War* (Marvel Wiki 2019). Redwing is flown using a body-worn touch screen, appears to use some type of mini-jet propulsion system, is highly maneuverable, and includes an onboard 360° camera that uses multiple spectrums (including X-ray!). Redwing provides real-time visuals to Falcon's goggles and includes an onboard grappling hook and two guns mounted on its wings. Redwing is also modular and able to split into multiple units while airborne.

Clearly, Redwing is a tactical drone built for military applications, but is it so ridiculous to envision a similarly capable μUAS for mapping and spatial analysis in urban areas? We think not. This may be a reality within the next decade, where urban planners, city managers, and various local agencies use fleets of μUAS for monitoring and mapping urban areas in near-real time. These μUAS will be equipped with a range of sensors for capturing information about urban areas. For example, one could use μUAS to respond to water main breaks to evaluate their severity before dispatching repair crews. However, rather than just a visual inspection from overhead, the micro drones may be able to access the inside of the water main itself, identifying additional areas of weakness that may benefit from a proactive replacement decision. A fleet of μUAS could be used to map and evaluate property (or perhaps urban infrastructure) damage after an extreme event. For example, after a hurricane, this near-real-time mapping could afford response agencies with enough geospatial intelligence to optimize the delivery of important relief supplies when local roads are flooded and impassible for most vehicles (Helderop and Grubesic 2019).

The real "killer" application for μUAS will be the inclusion of their data into the six core data streams that many of us are already familiar with: (1) social/entertainment, (2) Internet of things (IOT)/smart homes, (3) connected vehicles, (4) financial and enterprise, (5) online collaboration, and (6) audience participation (Table 10.1).[8] This may seem far-fetched, but consider what these different data streams are—bits

[7] Not to be confused with the Bengaluru-based startup, Redwing, which recently made a name for itself by building drones for delivering medical supplies throughout Papua New Guinea (Kashyap 2019).

[8] We assume that readers are casually familiar with many of the companies and platforms detailed in Table 10.1.

and pieces of information being produced and consumed by different devices and people. Information collected by a μUAS is nothing more than this. The question is, how does one meaningfully integrate this information into the six existing streams? Recent estimates suggest that there are more than five billion mobile devices operating in the world, with more than half consisting of smartphones. In other words, there is no shortage of devices that could consume data from μUAS or be used to control a μUAS. The current number of mobile devices is dwarfed by estimates concerning the number of devices that will be connected to the Internet by 2025, which is close to 34 billion (IOT Analytics 2019). Next, one needs an Internet-enabled sUAS or μUAS. Again, although the development of μUAS continues, Internet-connected sUAS already exist. For example, FlytCloud is a platform that helps manage UAS fleets for delivery, remote inspection, and security/surveillance. It facilitates real-time control and the reporting of flight data using 4G LTE networks on a web-based dashboard. More importantly, the FlytCloud application programming interface (API) is easily integrated via built.io into Dropbox, Cisco Webex, Slack, YouTube, Twitter, Instagram, and hundreds of other applications. In other words, this is already happening for sUAS.

Readers should also recognize that real-time mapping and analysis applications for UAS will consist of two-way data streams. For example, consider the strategic deployment of sUAS in regions recently struck by a hurricane. As the sUAS flies over a neighborhood, it may be collecting and reporting real-time imagery, but this imagery is being enriched off-site with important information about the people, places, and businesses within its survey area. This type of real-time geoenrichment includes demographics, socioeconomic status, and other relevant characteristics associated with the neighborhood (Esri 2019). Simultaneously, we anticipate that the sUAS fleets may serve as relays for post-disaster recovery efforts by retransmitting information from mobile devices concerning the general welfare of a neighborhood. This may include information on residents' health (pulled from health tracking software), social media posts, imagery collected at ground level, and other information that can be ingested and processed (or at least transmitted) by an Internet-enabled sUAS. In turn, all of this information can be used to provide a real-time snapshot of a neighborhood. More importantly, if enough capable sUAS or μUAS (in the future) are available, this can be done for an entire city or urban area.

10.4 Public Policy and Regulation

Before any of this happens, several significant changes need to be made to UAS policy and UAS regulation. First, regulators need to determine policies that allow for the widespread use of sUAS and μUAS, while still maintaining the security, safety, and privacy that society wants and expects. To some extent, this is underway at the United States Federal Aviation Administration (FAA). For example, in the United States, the FAA has exclusive authority to regulate aviation safety, which includes drones. As a result, cities and municipalities within the United States are not permitted to have their own rules or regulations governing the operation of UAS. This helps eliminate what would surely be a chaotic patchwork of local and state policies, many of which would likely stifle the use of UAS for research or commercial applications. That said, local municipalities can regulate where drones land, because this issue involves local control of land and zoning, which are not subject to federal regulation (FAA 2018).

Issues of personal privacy are likely to generate the most controversy when considering drone regulation. Not surprisingly, there is a growing body of research that attempts to address such issues (Scott 2017; Luppicini and So 2016; Yao et al. 2017; Li et al. 2019; Finn and Wright 2016), as well as strategies for improving privacy (Schaub and Knierim 2016; Sun et al. 2017). We recommend readers consult these sources (along with Chap. 5 of this book) to appreciate the variations in privacy concerns and policy variations recommended within. Looking forward, these issues will become more relevant as drones decrease in size, improve in capability, and become more widely used.

Within this context, issues of "mission creep" are of particular concern for drone watchdog groups. For example, in August 2014, the group "Drone Free LAPD" (DF-LAPD) sent a letter to Mayor Eric Garcetti asking for a meeting to discuss the use of drones for law enforcement in the city (DF-LAPD 2014). In particular, DF-LAPD is concerned that the application of a specific tactic, in this case drones, will expand beyond the original stated scope of use and move toward new and enlarged purposes. This may include the unlawful surveillance of individuals not involved in an investigation or incident, or the use of drones for license plate recognition to monitor the movements of private citizens. These are real concerns and something that federal, state, and local policymakers must grapple with as UAS technology becomes more mainstream in the future.

A more pragmatic concern is the growing need to operate UAS beyond visual line of site (BVLOS) and over people. The United States Department of Transportation recently issued some guidance on their efforts to more fully integrate UAS into the national air transport system. This includes information and regulation for sUAS activity above people and beyond visual line of sight (USDOT 2018). In short, this is happening, but the rollout will be slow. That said, the FAA recently issued the first certificates of authorization (COA) to fly BVLOS. The first certificate was issued to the University of Alaska—Fairbanks to monitor remote sections of the trans-Alaska pipeline (Moon 2019). In many respects, this is an excellent location to begin rolling out this type of waiver and testing BVLOS capabilities. Much of Alaska is sparsely populated so if the UAV encountered a problem and crashed, the chances of it hitting a person are low. We cannot say the same for a Caribou.

Interestingly, the second COA was issued to the Chula Vista police department (Solis 2019). In addition to this being one of the most densely populated places in the United States, issuing a COA to a police department is fascinating. Of course, Chula Vista PD is teaming up with a Cape Aerial Telepresence (https://www.cape.com/) for testing the operation of UAVs at extended visual lines of site (EVLOS) and BVLOS. This includes efforts to implement the appropriate technologies and protocols for UAVs to detect and avoid (DAA) collisions with active aircrafts. Efforts to manage spectrum allocation (i.e., radio frequencies used to fly UAVs) and technologies for real-time tracking of UAVs are also underway.

10.5 Conclusion

As Yoda once said, "Difficult to see. Always in motion is the future." This is both literally and figuratively true for UAS and the UAS industry, especially when it comes to mapping and spatial analysis. Although we made a few predictions, readers will recognize that we largely avoided any grand prognostications. We are fairly confident that the UAS of the future will be smaller, more efficient, and better at what they do. This is literally happening every day, but more work is required. We have no doubt that by the time this book is published, there will be significant advancements in UAS technology that we failed to cover. To be honest, it is almost impossible to keep up.

Where mapping and spatial analysis is concerned, the ability to collect, process, and analyze inexpensive, high-quality and easily accessible UAS-derived information represents an incredible opportunity for the environmental, physical, socioeconomic, and planning sciences. Although we only scratch the surface in this book, we hope that readers will be inspired by this content and pursue their own substantive passions. If, by chance, this leads to improvements in the quality of life for the residents of a community, facilitates efficient emergency responses after a disaster, or provides a way to save energy or sustain a sensitive habitat, then we consider this book a success. Lastly, we want to encourage all readers to fly their UAS with glee, but paying careful attention to the rules that make it possible for all of us to benefit from this technology. Have fun out there!

References

Armand, M., & Tarascon, J. M. (2008). Building better batteries. *Nature, 451*(7179), 652.

Ayers, M. S., & Reder, L. M. (1998). A theoretical review of the misinformation effect: Predictions from an activation-based memory model. *Psychonomic Bulletin & Review, 5*(1), 1–21.

Cherry, K. (2019). How cognitive biases influence how you think and act. *Verywell Mind*. Retrieved August 30, 2019, from https://tinyurl.com/yazwfd4r

DF-LAPD. (2014). Letter to Mayor Eric Garcetti. *Drone Free LAPD*. Retrieved August 13, 2019, from https://tinyurl.com/y5ydnu4p

Durrant-Whyte, H., & Bailey, T. (2006). Simultaneous localization and mapping: Part I. *IEEE Robotics & Automation Magazine, 13*(2), 99–110.

Esri. (2019). *Geoenrichment services*. Retrieved September 1, 2019, from https://tinyurl.com/y42mcfyd

FAA. (2018). *FAA statement – federal vs. local drone authority*. Retrieved August 12, 2019, from https://tinyurl.com/y6by9x46

Finn, R. L., & Wright, D. (2016). Privacy, data protection and ethics for civil drone practice: A survey of industry, regulators and civil society organisations. *Computer Law & Security Review, 32*(4), 577–586.

Fischhoff, B. (1975). Hindsight is not equal to foresight: The effect of outcome knowledge on judgment under uncertainty. *Journal of Experimental Psychology: Human Perception and Performance, 1*(3), 288.

Floreano, D., & Wood, R. J. (2015). Science, technology and the future of small autonomous drones. *Nature, 521*(7553), 460.

Furnham, A., & Boo, H. C. (2011). A literature review of the anchoring effect. *The Journal of Socio-Economics, 40*(1), 35–42.

Ha, P. (2010). Motorola DynaTAC 8000×. *Time*. Retrieved August 1, 2019, from https://tinyurl.com/mvd9u45

Helderop, E., & Grubesic, T. H. (2019). Flood evacuation and rescue: The identification of critical road segments using whole-landscape features. *Transportation Research Interdisciplinary Perspectives* https://doi.org/10.1016/j.trip.2019.100022

Intel. (2008). *Reducing Power with Hardware Accelerators*. Retrieved September 3, 2019, from https://tinyurl.com/wsujg6j

International Organization for Standardization [ISO]. (2012). *Robots and robotic devices — vocabulary ISO 8373:2012*. URL: https://www.iso.org/standard/55890.html

IOT Analytics. (2019). *State of the IOT, 2018*. Retrieved August 27, 2019, from https://tinyurl.com/y5cosjt4

Kashyap, S. (2019). Meet the 24-year-olds who have built drones that supply medicines and blood to Papua New Guinea. *YourStory*. Retrieved August 14, 2019, from https://tinyurl.com/yxhh8h7f

Li, Z., Gao, C., Yue, Q., & Fu, X. (2019). Toward drone privacy via regulating altitude and payload. In *International Conference on Computing, Networking and Communications (ICNC)* (pp. 562–566). IEEE.

Luppicini, R., & So, A. (2016). A technoethical review of commercial drone use in the context of governance, ethics, and privacy. *Technology in Society, 46*, 109–119.

Marvel Wiki (2019). *Redwing*. Retrieved September 2, 2019, from https://tinyurl.com/y6gqkqfo

Mohamed, N., Al-Jaroodi, J., Jawhar, I., Idries, A., & Mohammed, F. (2018). Unmanned aerial vehicles applications in future smart cities. *Technological Forecasting and Social Change*. https://doi.org/10.1016/j.techfore.2018.05.004

Moon, M. (2019). First FAA-approved beyond-line-of-sight drone flies in the US. *Engaget*. Retrieved from https://tinyurl.com/y3ot434m

Morgan, K. (2019). Why you're so bad at predicting the future. *Medium*. Retrieved September 5, 2019, from https://tinyurl.com/yxnmkgyj

NASA. (2019). *Luna 1*. Retrieved September 5, 2019, from https://tinyurl.com/y4x9wa59

Nickerson, R. S. (1998). Confirmation bias: A ubiquitous phenomenon in many guises. *Review of General Psychology, 2*(2), 175–220.

Ovtcharov, K., Ruwase, O., Kim, J. Y., Fowers, J., Strauss, K., & Chung, E. S. (2015). Accelerating deep convolutional neural networks using specialized hardware. *Microsoft Research Whitepaper, 2*(11), 1–4.

Peercy, P. S. (2000). The drive to miniaturization. *Nature, 406*(6799), 1023.

Saha, S. (2017). Meet Zanco Tiny T1, world's smallest phone. *India Today*. Retrieved September 15, 2019, from https://tinyurl.com/y2uxppyz

Schaub, F., & Knierim, P. (2016). Drone-based privacy interfaces: Opportunities and challenges. In *Twelfth Symposium on Usable Privacy and Security*. Denver: {USENIX}Association.

Scott, J. D. (2017). Drone surveillance: The FAA's obligation to respond to the privacy risks. *Fordham Urban Law Journal, 44*, 767.

Sensefly. (2019). *Sensefly Albris 4*. Retrieved from https://tinyurl.com/y6nnsaya

Solis, G. (2019). FAA allows Chula Vista to expand police drone program. *San Diego Union Tribune*. Retrieved September 5, 2019, from https://tinyurl.com/yy6m8huh

Sun, A., Gong, W., Shea, R., Liu, J., Liu, X. S., & Wang, Q. (2017). Drone privacy shield: A WiFi based defense. In *IEEE 28th Annual International Symposium on Personal, Indoor, and Mobile Radio Communications (PIMRC)* (pp. 1–5). IEEE.

US Department of Transportation [USDOT]. (2018). *Integration of civil Unmanned Aircraft Systems (UAS) in the National Airspace System (NAS) roadmap*. Retrieved from https://tinyurl.com/y4hugomk

Weinstein, N. D. (1982). Unrealistic optimism about susceptibility to health problems. *Journal of Behavioral Medicine, 5*(4), 441–460.

Weinstein, N. D. (1989). Optimistic biases about personal risks. *Science, 246*(4935), 1232–1234.

Yao, Y., Xia, H., Huang, Y., & Wang, Y. (2017). Free to fly in public spaces: Drone controllers' privacy perceptions and practices. In *Proceedings of the 2017 CHI Conference on Human Factors in Computing Systems* (pp. 6789–6793). ACM.

Index

T. H. Grubesic, J. R. Nelson, *UAVs and Urban Spatial Analysis*,
https://doi.org/10.1007/978-3-030-35865-5